GW00691815

DIE MULTIPLE UNITS

THIRTY-FIRST EDITION
2018

The complete guide to all Diesel Multiple Units and On-Track Machines which operate on the national railway network

Robert Pritchard & Peter Hall

ISBN 978 1909 431 42 3

CONTENTS

PROVISION OF INFORMATION

This book has been compiled with care to be as accurate as possible, but some information is not easily available and the publisher cannot be held responsible for any errors or omissions. We would like to thank the companies and individuals who have been helpful in supplying information to us. The authors of this series of books are always pleased to receive notification of any inaccuracies that may be found, to enhance future editions. Please send comments to:

Robert Pritchard, Platform 5 Publishing Ltd, 52 Broadfield Road, Sheffield, S8 0XJ, England.

e-mail: robert.pritchard@platform5.com   **Tel:** 0114 255 2625.

This book is updated to information received by 9 October 2017.

UPDATES

This book is updated to the Stock Changes given in **Today's Railways UK 191** (November 2017). The Platform 5 railway magazine "**Today's Railways UK**" publishes Stock Changes every month to update this book. The magazine also contains news and rolling stock information on the railways of Great Britain and Ireland and is published on the second Monday of every month. For further details of **Today's Railways UK**, please contact Platform 5 Publishing Ltd.

Front cover photograph: Grand Central-liveried 180 114 passes Grantham with the 12.12 Sunderland–London King's Cross on 23/04/17.   **Robert Pritchard**

BRITAIN'S RAILWAY SYSTEM

INFRASTRUCTURE & OPERATION

Britain's national railway infrastructure is owned by a "not for dividend" company, Network Rail. In 2014 Network Rail was reclassified as a public sector company, being described by the Government as a "public sector arm's-length body of the Department for Transport".

Most stations and maintenance depots are leased to and operated by Train Operating Companies (TOCs), but some larger stations are controlled by Network Rail. The only exception is the infrastructure on the Isle of Wight: The Island Line franchise uniquely included maintenance of the infrastructure as well as the operation of passenger services. As Island Line is now part of the South West Trains franchise, both the infrastructure and trains are operated by South West Trains.

Trains are operated by TOCs over Network Rail tracks (the National Network), regulated by access agreements between the parties involved. In general, TOCs are responsible for the provision and maintenance of the locomotives, rolling stock and staff necessary for the direct operation of services, whilst Network Rail is responsible for the provision and maintenance of the infrastructure and also for staff to regulate the operation of services.

The Department for Transport (DfT) is the franchising authority for the national network, with Transport Scotland overseeing the award of the ScotRail franchise and the Welsh Government overseeing the Wales & Borders franchise.

A franchise is the right to run specified services within a specified area for a period of time, in return for the right to charge fares and, where appropriate, to receive financial support from the Government. Subsidy is payable in respect of socially necessary services. Service standards are monitored by the DfT throughout the duration of the franchise. Franchisees earn revenue primarily from fares and from subsidy. They generally lease stations from Network Rail and earn rental income by sub-letting parts of them, for example to retailers.

Franchisees' main costs are the track access charges they pay to Network Rail, the costs of leasing stations and rolling stock and of employing staff. Franchisees may do light maintenance work on rolling stock or contract it out to private companies. Heavy maintenance is normally carried out by the Rolling Stock Leasing Companies, according to contracts.

TOCs can take commercial risks, although some franchises are "management contracts", where ticket revenues pass directly to the DfT. Concessions (such as London Overground) see the operator paid a fee to run the service, usually within tightly specified guidelines. Operators running a concession would not normally take commercial risks, although there are usually penalties and rewards in the contract.

Note that a railway "reporting period" is four weeks.

DOMESTIC PASSENGER TRAIN OPERATORS

The majority of passenger trains are operated by Train Operating Companies on fixed-term franchises or concessions. Expiry dates are shown in the list below:

Franchise	Franchisee	Trading Name
Caledonian Sleeper	Serco	**Caledonian Sleeper**
	(until 31 March 2030)	

This new franchise started in April 2015 when operation of the ScotRail and ScotRail Sleeper franchises was separated. Abellio won the ScotRail franchise and Serco the Caledonian Sleeper franchise. Caledonian Sleeper operates four trains nightly between London Euston and Scotland using locomotives hired from GBRf or Freightliner. New CAF rolling stock will be introduced from spring 2018 to replace the current Mark 2 and Mark 3 carriages that are used.

Chiltern	Arriva (Deutsche Bahn)	**Chiltern Railways**
	(until 31 December 2021)	

There is an option to extend the franchise by 7 months to July 2022.

Chiltern Railways operates a frequent service between London Marylebone, Banbury and Birmingham Snow Hill, with some peak trains extending to Kidderminster. There are also regular services from Marylebone to Stratford-upon-Avon and to Aylesbury Vale Parkway via Amersham (along the London Underground Metropolitan Line). A new route to Oxford Parkway and then Oxford was added in 2015–16. The fleet consists of DMUs of Classes 121 (used on the Princes Risborough–Aylesbury route), 165, 168 and 172 plus a number of locomotive-hauled rakes used on some of the Birmingham route trains, worked by Class 68s hired from DRS.

Cross Country	Arriva (Deutsche Bahn)	**CrossCountry**
	(until December 2019)	

There is an option to extend the franchise by 11 months to November 2020.

CrossCountry operates a network of long distance services between Scotland, the North-East of England and Manchester to the South-West of England, Reading, Southampton, Bournemouth and Guildford, centred on Birmingham New Street. These trains are mainly formed of diesel Class 220/221 Voyagers, supplemented by a small number of HSTs on the NE–SW route. Inter-urban services also link Nottingham, Leicester and Stansted Airport with Birmingham and Cardiff. These trains use Class 170 DMUs.

Crossrail	MTR	**TfL Rail**
	(until 30 May 2023)	

There is an option to extend the concession by 2 years to May 2025.

This is a new concession which started in May 2015. Initially Crossrail took over the Liverpool Street–Shenfield stopping service from Greater Anglia, using a fleet of Class 315 EMUs, with the service branded "TfL Rail". New Class 345 EMUs will be introduced on this route from May 2017 and then from 2018–19 Crossrail will operate through new tunnels beneath central London, from Shenfield and Abbey Wood in the east to Reading and Heathrow Airport in the west. It will then be branded "Elizabeth Line".

East Coast	Stagecoach/Virgin Trains	**Virgin Trains East Coast**
	(until 31 March 2023)	

There is an option to extend the franchise by 1 year to March 2024.

Virgin Trains East Coast operates frequent long distance trains on the East Coast Main Line between London King's Cross, Leeds, York, Newcastle and Edinburgh, with less frequent services to Bradford, Harrogate, Skipton, Hull, Lincoln, Glasgow, Aberdeen and Inverness. A mixed fleet of Class 91s and 30 Mark 4 sets, and 15 HST sets, are used on these trains.

| **East Midlands** | Stagecoach Group (until March 2019) | **East Midlands Trains** |

There is an option to extend the franchise until August 2019, with a further option to extend it to August 2020.

EMT operates a mix of long distance high speed services on the Midland Main Line (MML), from London St Pancras to Sheffield (to Leeds at peak times and with some extensions to York/Scarborough) and Nottingham (plus peak-hour trains to Lincoln), and local and regional services ranging from the Norwich–Liverpool route to Nottingham–Skegness, Newark–Mansfield–Worksop, Nottingham–Matlock and Derby–Crewe. It also operates local services in Lincolnshire. Trains on the MML are worked by a fleet of Class 222 DMUs and nine HSTs, whilst the local and regional fleet consists of DMU Classes 153, 156 and 158.

| **East Anglia** | Abellio (Netherlands Railways)/Mitsui Group (until 15 October 2025) | **Greater Anglia** |

There is an option to extend the franchise by 1 year to October 2026.

Greater Anglia operates main line trains between London Liverpool Street, Ipswich and Norwich and local trains across Norfolk, Suffolk and parts of Cambridgeshire. It also runs local and commuter services into Liverpool Street from the Great Eastern (including Southend, Braintree and Clacton) and West Anglia (including Cambridge and Stansted Airport) routes. It operates a varied fleet of Class 90s with locomotive-hauled Mark 3 sets, DMUs of Classes 153, 156 and 170 and EMUs of Classes 317, 321, 360 and 379. One locomotive-hauled set, using Class 37s, is currently hired for use on some trains between Norwich and Great Yarmouth/Lowestoft.

| **Essex Thameside** | Trenitalia (until 8 November 2029) | **c2c** |

There is an option to extend the franchise by 6 months to May 2030.

c2c operates an intensive, principally commuter, service from London Fenchurch Street to Southend and Shoeburyness via both Upminster and Tilbury. The fleet consists of 74 Class 357 EMUs, plus six Class 387s which arrived in late 2016. In 2014 c2c won the new 15-year franchise that promised to introduce 17 new 4-car EMUs from 2019.

| **Great Western** | First Group (until 1 April 2020) | **Great Western Railway** |

Great Western Railway (until September 2015 branded as First Great Western) operates long distance trains from London Paddington to South Wales, the West Country and Worcester and Hereford. In addition there are frequent trains along the Thames Valley corridor to Newbury/Bedwyn and Oxford, plus local and regional trains throughout the South-West including the Cornish, Devon and Thames Valley branches, the Reading–Gatwick North Downs Line and Cardiff–Portsmouth Harbour and Bristol–Weymouth regional routes. A large fleet of HSTs is used on the long-distance trains, with DMUs of Classes 165 and 166 used on the Thames Valley and North Downs routes and Class 180s used alongside HSTs on the Cotswold Line to Worcester and Hereford. New Class 387 EMUs are now also being used on initial electric services from Paddington to Maidenhead. Classes 143, 150, 153 and 158 are used on local and regional trains in the South-West. A small fleet of Class 57s is maintained to principally work the overnight "Cornish Riviera" Sleeper service between London Paddington and Penzance. Class 800 IET units will be introduced in autumn 2017 to replace the HSTs and Class 180s.

| **London Rail** | Arriva (Deutsche Bahn) (until May 2024) | **London Overground** |

This is a concession and is different from other rail franchises, as fares and service levels are set by Transport for London instead of by the DfT. There is an option to extend the concession by 2 years to May 2026.

London Overground operates services on the Richmond–Stratford North London Line and the Willesden Junction–Clapham Junction West London Line, plus the East London Line from Highbury & Islington to New Cross and New Cross Gate, with extensions to Clapham Junction (via Denmark Hill), Crystal Palace and West Croydon. It also runs services from London Euston to Watford Junction. All these use Class 378 EMUs whilst Class 172 DMUs are used on the Gospel Oak–Barking route. London Overground also took over the operation of some suburban services from London Liverpool Street in 2015 – to Chingford, Enfield Town and Cheshunt. These use Class 315 and 317 EMUs, but will be replaced by new Bombardier Class 710s from 2018 – these will also be used on the Gospel Oak–Barking line, which is being electrified.

Merseyrail Electrics   Serco/Abellio (Netherlands Railways)   **Merseyrail**
                                        (until 19 July 2028)
Under the control of Merseytravel PTE instead of the DfT. Franchise reviewed every five years to fit in with the Merseyside Local Transport Plan.

Merseyrail operates services between Liverpool and Southport, Ormskirk, Kirkby, Hunts Cross, New Brighton, West Kirby, Chester and Ellesmere Port, using Class 507 and 508 EMUs.

Northern   Arriva (Deutsche Bahn)   **Northern**
                                        (until 31 March 2025)
There is an option to extend the franchise by 1 year to March 2026.

Northern operates a range of inter-urban, commuter and rural services throughout the North of England, including those around the cities of Leeds, Manchester, Sheffield, Liverpool and Newcastle. The network extends from Chathill in the north to Nottingham in the south, and Cleethorpes in the east to St Bees in the west. Long distance services include Leeds–Carlisle, Middlesbrough–Carlisle and York–Blackpool North. The operator uses a large fleet of DMUs of Classes 142, 144, 150, 153, 155, 156 and 158 plus EMU Classes 319, 321, 322, 323 and 333. Class 185s are hired from TransPennine Express for use on some services between Manchester Airport and Blackpool North/Barrow/Windermere. Two locomotive-hauled sets, with Class 37s, are hired from DRS for use on some trains between Carlisle and Barrow-in-Furness/Preston as part of a 4-year contract that started in May 2015.

ScotRail   Abellio (Netherlands Railways)   **ScotRail**
                                        (until 31 March 2022)
There is an option to extend the franchise by 3 years to March 2025.

ScotRail provides almost all passenger services within Scotland and also trains from Glasgow to Carlisle via Dumfries, some of which extend to Newcastle (jointly operated with Northern). The company operates a large fleet of DMUs of Classes 156, 158 and 170 and EMU Classes 314, 318, 320, 334 and 380. Two locomotive-hauled rakes are also used on Fife Circle commuter trains, hauled by Class 68s hired from DRS.

South Eastern   Govia (Go-Ahead/Keolis)   **Southeastern**
                                        (until 24 December 2018)
There is an option to extend the franchise by a further 4 weeks to January 2019.

Southeastern operates all services in the South-East London suburbs, the whole of Kent and part of Sussex, which are primarily commuter services to London. It also operates domestic high speed trains on HS1 from London St Pancras to Ashford, Ramsgate, Dover and Faversham with additional peak services on other routes. EMUs of Classes 375, 376, 377, 465 and 466 are used, along with Class 395s on the High Speed trains.

| South Western | First Group/MTR | **South Western Railway** |
| | (until 18 August 2024) | |

There is an option to extend the franchise by 11 months to July 2025.

South Western Railway operates trains from London Waterloo to destinations across the South and South-West including Woking, Basingstoke, Southampton, Portsmouth, Salisbury, Exeter, Reading and Weymouth as well as suburban services from Waterloo. SWR also runs services between Ryde and Shanklin on the Isle of Wight, using former London Underground 1938 stock (Class 483s). The rest of the fleet consists of DMU Classes 158 and 159 and EMU Classes 444, 450, 455, 456, 458 and 707.

| **Thameslink, Southern &** | Govia (Go-Ahead/Keolis) | **Govia Thameslink Railway** |
| **Great Northern (TSGN)** | (until 19 September 2021) | |

There is an option to extend the franchise by 2 years to September 2023.

Govia operates this franchise, the largest in the UK, as a management contract. The former Southern franchise was combined with Thameslink/Great Northern in 2015. GTR uses four brands within the franchise: "Thameslink" for trains between Bedford and Brighton via central London and also on the Sutton/Wimbledon loop using new Class 700 EMUs. Some trains to Sevenoaks are jointly operated with Southeastern. "Great Northern" comprises services from London King's Cross and Moorgate to Welwyn Garden City, Hertford North, Peterborough, Cambridge and Kings Lynn using Class 313, 365 and 387 EMUs. "Southern" operates predominantly commuter services between London, Surrey and Sussex and "metro" services in South London, as well as services along the South Coast between Southampton, Brighton, Hastings and Ashford, plus the cross-London service from South Croydon to Milton Keynes. Class 171 DMUs are used on Brighton–Ashford and London Bridge–Uckfield services, whilst all other services are in the hands of Class 313, 377, 455 and 700 EMUs. Finally, the premium "Gatwick Express" operates non-stop trains between London Victoria, Gatwick Airport and Brighton using Class 387/2 EMUs.

| **Trans-Pennine Express** | First Group | **TransPennine Express** |
| | (until 31 March 2023) | |

There is an option to extend the franchise by 2 years to March 2025.

TransPennine Express operates predominantly long distance inter-urban services linking major cities across the North of England, along with Edinburgh and Glasgow in Scotland. The main services are Manchester Airport/Manchester Piccadilly–Newcastle/Middlesbrough/Hull plus Liverpool–Scarborough and Liverpool–Newcastle along the North Trans-Pennine route via Huddersfield, Leeds and York, and Manchester Airport–Cleethorpes along the South Trans-Pennine route via Sheffield. TPE also operates Manchester Airport–Edinburgh/Glasgow. The fleet consists of Class 185 DMUs, plus Class 350 EMUs used on Manchester Airport–Scotland services.

| **Wales & Borders** | Arriva (Deutsche Bahn) | **Arriva Trains Wales** |
| | (until 14 October 2018) | |

There is an option to extend the franchise by 6 months to April 2019. The Welsh Government will be procuring the next Wales & Borders franchise.

Arriva Trains Wales operates a mix of long distance, regional and local services throughout Wales, including the Valley Lines network of lines around Cardiff, and also through services to the English border counties and to Manchester and Birmingham. The fleet consists of DMUs of Classes 142, 143, 150, 158 and 175 and two locomotive-hauled rakes: one used on a premium Welsh Government sponsored service on the Cardiff–Holyhead route, and one used between Manchester/Crewe and Holyhead (both are hauled by a Class 67).

West Coast Partnership   Virgin Rail Group (Virgin/Stagecoach Group)   **Virgin Trains**
                                  (until 31 March 2019)

Virgin Trains operates long distance services along the West Coast Main Line from London Euston to Birmingham/Wolverhampton, Manchester, Liverpool and Glasgow using Class 390 Pendolino EMUs. It also operates Class 221 Voyagers on the Euston–Chester–Holyhead route, whilst a mixture of Class 221s and 390s are used on the Euston–Birmingham–Glasgow/Edinburgh route.

West Midlands   Govia (Go-Ahead/Keolis)   **London Midland**
                                 (until 10 December 2017)

A new franchise starts on 11 December 2017 and will be operated by a consortium of Abellio/ East Japan Railway Company/Mitsui.

London Midland operates long distance and regional services from London Euston to Northampton and Birmingham/Crewe and also between Birmingham and Liverpool as well as local and regional services around Birmingham, including to Stratford-upon-Avon, Worcester, Hereford, Redditch and Shrewsbury. It also operates the Bedford–Bletchley and Watford Junction–St Albans Abbey branches. The fleet consists of DMU Classes 150, 153, 170 and 172 and EMU Classes 319, 323 and 350.

NON-FRANCHISED SERVICES

The following operators run non-franchised, or "open access" services:

Operator	Trading Name	Route
Heathrow Airport Holdings	Heathrow Express	London Paddington–Heathrow Airport
Hull Trains (part of First)	Hull Trains	London King's Cross–Hull
Grand Central (part of Arriva)	Grand Central	London King's Cross–Sunderland/ Bradford Interchange
North Yorkshire Moors Railway Enterprises	North Yorkshire Moors Railway	Pickering–Grosmont–Whitby/ Battersby, Sheringham–Cromer
West Coast Railway Company	West Coast Railway Company	Birmingham–Stratford-upon-Avon Fort William–Mallaig York–Settle–Carlisle

INTERNATIONAL PASSENGER OPERATORS

Eurostar International operates passenger services between the UK and mainland Europe. The company, established in 2010, is jointly owned by SNCF (the national operator of France): 55%, SNCB (the national operator of Belgium): 5% and Patina Rail: 40%. Patina Rail is made up of Canadian-based Caisse de dépôt et placement du Québec (CDPG) and UK-based Hermes Infrastructure (owning 30% and 10% respectively). This 40% was previously owned by the UK Government until it was sold in 2015.

In addition, a service for the conveyance of accompanied road vehicles through the Channel Tunnel is provided by the tunnel operating company, Eurotunnel. All Eurotunnel services are operated in top-and-tail mode by the powerful Class 9 Bo-Bo-Bo locomotives.

INTRODUCTION

This book contains details of all Diesel Multiple Units, usually referred to as DMUs, which can run on Britain's national railway network.

Since the 1980s DMUs have replaced more traditional locomotive-hauled trains on many routes. DMUs today work a wide variety of services, from long distance Intercity to inter-urban and suburban duties.

LAYOUT OF INFORMATION

DMUs are listed in numerical order of set – using current numbers as allocated by the RSL. Individual "loose" vehicles are listed in numerical order after vehicles formed into fixed formations. Where sets or vehicles have been renumbered in recent years, former numbering detail is shown in parentheses. Each entry is laid out as in the following example:

RSL Set No.	Detail	Livery	Owner	Operator	Depot	Formation	Name
153 309	cr	**GA**	P	*GA*	NC	52309	GERALD FIENNES

Codes: Codes are used to denote the livery, owner, operator and depot allocation of each Diesel Multiple Unit. Details of these can be found in section 6 of this book. Where a unit or spare car is off-lease, the operator column is left blank.

Detail Differences: Detail differences which currently affect the areas and types of train which vehicles may work are shown, plus differences in interior layout. Where such differences occur within a class, these are shown either in the heading information or alongside the individual set or vehicle number. The following standard abbreviations are used:

e   European Railway Traffic Management System (ERTMS) signalling equipment fitted.
r   Radio Electric Token Block signalling equipment fitted.

Use of the above abbreviations indicates the equipment fitted is normally operable. Meaning of non-standard abbreviations is detailed in individual class headings.

Set Formations: Regular set formations are shown where these are normally maintained. Readers should note set formations might be temporarily varied from time to time to suit maintenance and/or operational requirements. Vehicles shown as "Spare" are not formed in any regular set formation.

Names: Only names carried with official sanction are listed. Names are shown in UPPER/lower case characters as actually shown on the name carried on the vehicle(s). Unless otherwise shown, complete units are regarded as named rather than just the individual car(s) which carry the name.

GENERAL INFORMATION

CLASSIFICATION AND NUMBERING

DMU Classes are listed in class number order.

First generation ("Heritage") DMUs were classified in the series 100–139.
Parry People Movers (not technically DMUs) are classified in the series 139.
Second generation DMUs are classified in the series 140–199.
Diesel Electric Multiple Units are classified in the series 200–249.
Service units are classified in the series 930–999.

First and second generation individual cars are numbered in the series 50000–59999 and 79000–79999.

Parry People Mover cars are numbered in the 39000 series.

DEMU individual cars are numbered in the series 60000–60999, except for a few former EMU vehicles which retain their EMU numbers.

For all new vehicles allocated by the Rolling Stock Library since 2014 6-digit vehicle numbers are being used. The Class 230 D-Train DEMU individual cars are numbered in the 300xxx series.

Service Stock individual cars are numbered in the series 975000–975999 and 977000–977999, although this series is not exclusively used for DMU vehicles.

WHEEL ARRANGEMENT

A system whereby the number of powered axles on a bogie or frame is denoted by a letter (A = 1, B = 2, C= 3 etc) and the number of unpowered axles is denoted by a number is used in this publication. The letter "o" after a letter indicates that each axle is individually powered.

UNITS OF MEASUREMENT

Principal details and dimensions are quoted for each class in metric and/or imperial units as considered appropriate bearing in mind common UK usage.

All dimensions and weights are quoted for vehicles in an "as new" condition with all necessary supplies (eg oil, water, sand) on board. Dimensions are quoted in the order Length – Width. All lengths quoted are over buffers or couplers as appropriate. Where two lengths are quoted, the first refers to outer vehicles in a set and the second to inner vehicles. All width dimensions quoted are maxima. All weights are shown as metric tonnes (t = tonnes).

OPERATING CODES

These codes are used by train operating company staff to describe the various different types of vehicles and normally appear on data panels on the inner (ie non driving) ends of vehicles.

The first part of the code describes whether the car has a motor or a driving cab as follows:

DM Driving motor   DT Driving trailer   M Motor   T Trailer

The next letter is a "B" for cars with a brake compartment.
This is followed by the saloon details:

F	First	L	denotes a vehicle with a toilet.
S	Standard	W	denotes a Wheelchair space.
C	Composite		

Finally vehicles with a buffet or kitchen area are suffixed RB or RMB for a miniature buffet counter.

Where two vehicles of the same type are formed within the same unit, the above codes may be suffixed by (A) and (B) to differentiate between the vehicles.

A composite is a vehicle containing both First and Standard Class accommodation, whilst a brake vehicle is a vehicle containing separate specific accommodation for the conductor.

Where vehicles have been declassified, the correct operating code which describes the actual vehicle layout is quoted in this publication.

BUILD DETAILS

Vehicles ordered under the auspices of BR were allocated a Lot (batch) number when ordered and these are quoted in class headings and sub-headings. Vehicles ordered since 1995 have no Lot Numbers, but the manufacturer and location that they were built is given.

ACCOMMODATION

The information given in class headings and sub-headings is in the form F/S nT (or TD) nW. For example 12/54 1T 1W denotes 12 First Class and 54 Standard Class seats, one toilet and one space for a wheelchair. A number in brackets (ie (+2)) denotes tip-up seats (in addition to the fixed seats). Tip-up seats in vestibules do not count. The seating layout of open saloons is shown as 2+1, 2+2 or 3+2 as the case may be. Where units have First Class accommodation as well as Standard Class and the layout is different for each class then these are shown separately prefixed by "1:" and "2:". TD denotes a universal access toilet suitable for use by a disabled person.

ABBREVIATIONS

The following abbreviations are used throughout this publication:

BR	British Railways.	km/h	kilometres per hour.
BSI	Bergische Stahl Industrie.	kW	kilowatts.
DEMU	Diesel Electric Multiple Unit.	LT	London Transport.
DMU	Diesel Multiple Unit .	LUL	London Underground Limited.
EMU	Electric Multiple Unit.	m	metres.
kN	kilonewtons.	mph	miles per hour.

1. DIESEL MECHANICAL & DIESEL HYDRAULIC UNITS

1.1. PARRY PEOPLE MOVERS

CLASS 139   PPM-60

Gas/flywheel hybrid drive Railcars used on the Stourbridge Junction–Stourbridge Town branch.
Body construction: Stainless steel framework.
Chassis construction: Welded mild steel box section.
Primary Drive: Ford MVH420 2.3 litre 64 kW (86 hp) LPG fuel engine driving through Newage marine gearbox, Tandler bevel box and 4 "V" belt driver to flywheel.
Flywheel Energy Store: 500 kg, 1 m diameter, normal operational speed range 1000–1500 rpm.
Final transmission: 4 "V" belt driver from flywheel to Tandler bevel box, Linde hydrostatic transmission and spiral bevel gearbox at No. 2 end axle.
Braking: Normal service braking by regeneration to flywheel (1 m/s/s); emergency/parking braking by sprung-on, air-off disc brakes (3 m/s/s).
Maximum Speed: 45 mph.
Dimensions: 8.7 x 2.4 m.
Doors: Deans powered doors, double-leaf folding (one per side).
Seating Layout: 1+1 unidirectional/facing.
Multiple Working: Not applicable.

39001–002. DMS. Main Road Sheet Metal, Leyland 2007–08. –/17(+4) 1W. 12.5 t.

139 001	**LM**	P	*LM*	SJ	39001
139 002	**LM**	P	*LM*	SJ	39002

1.2. SECOND GENERATION UNITS

All units in this section have air brakes and are equipped with public address, with transmission equipment on driving vehicles and flexible diaphragm gangways. Except where otherwise stated, transmission is Voith 211r hydraulic with a cardan shaft to a Gmeinder GM190 final drive.

CLASS 142   PACER   BREL DERBY/LEYLAND

DMS–DMSL.

Construction: Steel underframe, rivetted steel body and roof. Built from Leyland National bus parts on Leyland Bus four-wheeled underframes.
Engines: One Cummins LT10-R of 165 kW (225 hp) at 1950 rpm.
Couplers: BSI at outer ends, bar within unit.
Dimensions: 15.55 x 2.80 m.
Gangways: Within unit only.   **Wheel Arrangement:** 1-A + A-1.
Doors: Twin-leaf inward pivoting.   **Maximum Speed:** 75 mph.
Seating Layout: 3+2 mainly unidirectional bus/bench style unless stated.
Multiple Working: Within class and with Classes 143, 144, 150, 153, 155, 156, 158 and 159.

c   Refurbished Arriva Trains Wales units. Fitted with 2+2 individual Chapman seating.
s   Fitted with 2+2 individual high-back seating.
t   Former First North Western facelifted units – DMS fitted with a luggage/bicycle rack and wheelchair space.
u   Merseytravel units – Fitted with 3+2 individual low-back seating.

55542–591. DMS. Lot No. 31003 1985–86. –/62 (c –/46(+6) 2W, s –/56, t –/53 or 55 1W, u –/52 or 54 1W). 24.5 t.
55592–641. DMSL. Lot No. 31004 1985–86. –/59 1T (c –/44(+6) 1T 2W, s –/50 1T, u –/49 1T). 25.0 t.
55701–746. DMS. Lot No. 31013 1986–87. –/62 (c –/46(+6) 2W, s –/56, t –/53 or 55 1W, u –/52 or 54 1W). 24.5 t.
55747–792. DMSL. Lot No. 31014 1986–87. –/59 1T (c –/44(+6) 1T 2W, s –/50 1T, u –/60 1T). 25.0 t.

142 001	t	**NO**	A	*NO*	NH	55542	55592
142 002	c	**AV**	A	*AW*	CF	55543	55593
142 003		**NO**	A	*NO*	NH	55544	55594
142 004	t	**NO**	A	*NO*	NH	55545	55595
142 005	t	**NO**	A	*NO*	NH	55546	55596
142 006	c	**AV**	A	*AW*	CF	55547	55597
142 007	t	**NO**	A	*NO*	NH	55548	55598
142 009	t	**NO**	A	*NO*	HT	55550	55600
142 010	c	**AV**	A	*AW*	CF	55551	55601
142 011	t	**NO**	A	*NO*	NH	55552	55602
142 012	t	**NO**	A	*NO*	NH	55553	55603
142 013		**NO**	A	*NO*	NH	55554	55604
142 014	t	**NO**	A	*NO*	NH	55555	55605
142 015	s	**NO**	A	*NO*	HT	55556	55606

142 016	s		**N0**	A	*N0*	HT	55557	55607
142 017	s		**N0**	A	*N0*	HT	55558	55608
142 018	s		**N0**	A	*N0*	HT	55559	55609
142 019	s		**N0**	A	*N0*	HT	55560	55610
142 020	s		**N0**	A	*N0*	HT	55561	55611
142 021	s		**N0**	A	*N0*	HT	55562	55612
142 022	s		**N0**	A	*N0*	HT	55563	55613
142 023	t		**N0**	A	*N0*	HT	55564	55614
142 024	s		**N0**	A	*N0*	HT	55565	55615
142 025	s		**N0**	A	*N0*	HT	55566	55616
142 026	s		**N0**	A	*N0*	HT	55567	55617
142 027	t		**N0**	A	*N0*	NH	55568	55618
142 028	t		**N0**	A	*N0*	NH	55569	55619
142 029			**N0**	A	*N0*	NH	55570	55620
142 030			**N0**	A	*N0*	NH	55571	55621
142 031	t		**N0**	A	*N0*	NH	55572	55622
142 032	t		**N0**	A	*N0*	NH	55573	55623
142 033	t		**N0**	A	*N0*	NH	55574	55624
142 034	t		**N0**	A	*N0*	NH	55575	55625
142 035	t		**N0**	A	*N0*	NH	55576	55626
142 036	t		**N0**	A	*N0*	NH	55577	55627
142 037	t		**N0**	A	*N0*	NH	55578	55628
142 038	t		**N0**	A	*N0*	NH	55579	55629
142 039	t		**N0**	A	*N0*	NH	55580	55630
142 040	t		**N0**	A	*N0*	NH	55581	55631
142 041	u		**N0**	A	*N0*	NH	55582	55632
142 042	u		**N0**	A	*N0*	NH	55583	55633
142 043	u		**N0**	A	*N0*	NH	55584	55634
142 044	u		**N0**	A	*N0*	NH	55585	55635
142 045	u		**N0**	A	*N0*	NH	55586	55636
142 046	u		**N0**	A	*N0*	NH	55587	55637
142 047	u		**N0**	A	*N0*	NH	55588	55638
142 048	u		**N0**	A	*N0*	NH	55589	55639
142 049	u		**N0**	A	*N0*	NH	55590	55640
142 050	s		**N0**	A	*N0*	HT	55591	55641
142 051	u		**N0**	A	*N0*	NH	55701	55747
142 052	u		**N0**	A	*N0*	NH	55702	55748
142 053	u		**N0**	A	*N0*	NH	55703	55749
142 054	u		**N0**	A	*N0*	NH	55704	55750
142 055	u		**N0**	A	*N0*	NH	55705	55751
142 056	u		**N0**	A	*N0*	NH	55706	55752
142 057	u		**N0**	A	*N0*	NH	55707	55753
142 058	u		**N0**	A	*N0*	NH	55708	55754
142 060	t		**N0**	A	*N0*	NH	55710	55756
142 061	t		**N0**	A	*N0*	NH	55711	55757
142 062	t		**N0**	A	*N0*	NH	55712	55758
142 063	t		**N0**	A	*N0*	NH	55713	55759
142 064	t		**N0**	A	*N0*	HT	55714	55760
142 065	s		**N0**	A	*N0*	HT	55715	55761
142 066	s		**N0**	A	*N0*	HT	55716	55762
142 067			**N0**	A	*N0*	HT	55717	55763

142 068	t	**NO**	A	*NO*	HT	55718 55764
142 069	c	**AV**	A	*AW*	CF	55719 55765
142 070	t	**NO**	A	*NO*	HT	55720 55766
142 071	s	**NO**	A	*NO*	HT	55721 55767
142 072	c	**AV**	A	*AW*	CF	55722 55768
142 073	c	**AV**	A	*AW*	CF	55723 55769
142 074	c	**AV**	A	*AW*	CF	55724 55770
142 075	c	**AV**	A	*AW*	CF	55725 55771
142 076	c	**AV**	A	*AW*	CF	55726 55772
142 077	c	**AV**	A	*AW*	CF	55727 55773
142 078	s	**NO**	A	*NO*	HT	55728 55774
142 079	s	**NO**	A	*NO*	HT	55729 55775
142 080	c	**AV**	A	*AW*	CF	55730 55776
142 081	c	**AV**	A	*AW*	CF	55731 55777
142 082	c	**AV**	A	*AW*	CF	55732 55778
142 083	c	**AV**	A	*AW*	CF	55733 55779
142 084	s	**NO**	A	*NO*	HT	55734 55780
142 085	c	**AV**	A	*AW*	CF	55735 55781
142 086	s	**NO**	A	*NO*	HT	55736 55782
142 087	s	**NO**	A	*NO*	HT	55737 55783
142 088	s	**NO**	A	*NO*	HT	55738 55784
142 089	s	**NO**	A	*NO*	HT	55739 55785
142 090	s	**NO**	A	*NO*	HT	55740 55786
142 091	s	**NO**	A	*NO*	HT	55741 55787
142 092	s	**NO**	A	*NO*	HT	55742 55788
142 093	s	**NO**	A	*NO*	HT	55743 55789
142 094	s	**NO**	A	*NO*	HT	55744 55790
142 095	s	**NO**	A	*NO*	HT	55745 55791
142 096	s	**NO**	A	*NO*	HT	55746 55792

CLASS 143   PACER   ALEXANDER/BARCLAY

DMS–DMSL. Similar design to Class 142, but bodies built by W Alexander with Barclay underframes.

Construction: Steel underframe, aluminium alloy body and roof. Alexander bus bodywork on four-wheeled underframes.
Engines: One Cummins LT10-R of 165 kW (225 hp) at 1950 rpm.
Couplers: BSI at outer ends, bar within unit.
Dimensions: 15.45 x 2.80 m.
Gangways: Within unit only.   **Wheel Arrangement**: 1-A + A-1.
Doors: Twin-leaf inward pivoting.   **Maximum Speed**: 75 mph.
Seating Layout: 2+2 high-back Chapman seating, mainly unidirectional.
Multiple Working: Within class and with Classes 142, 144, 150, 153, 155, 156, 158 and 159.

DMS. Lot No. 31005 Andrew Barclay 1985–86. –/48(+6) 2W. 24.0 t.
DMSL. Lot No. 31006 Andrew Barclay 1985–86. –/44(+6) 1T 2W. 24.5 t.

143 601	**AV**	MG	*AW*	CF	55642 55667
143 602	**AW**	P	*AW*	CF	55651 55668
143 603	**GW**	P	*GW*	EX	55658 55669

143 604	**AW**	P	*AW*	CF	55645	55670	
143 605	**AW**	P	*AW*	CF	55646	55671	
143 606	**AW**	P	*AW*	CF	55647	55672	
143 607	**AW**	P	*AW*	CF	55648	55673	
143 608	**AW**	P	*AW*	CF	55649	55674	
143 609	**AV**	SG	*AW*	CF	55650	55675	Sir Tom Jones
143 610	**AV**	MG	*AW*	CF	55643	55676	
143 611	**GW**	P	*GW*	EX	55652	55677	
143 612	**GW**	P	*GW*	EX	55653	55678	
143 614	**AV**	MG	*AW*	CF	55655	55680	
143 616	**AW**	P	*AW*	CF	55657	55682	
143 617	**GW**	GW	*GW*	EX	55644	55683	
143 618	**FI**	GW	*GW*	EX	55659	55684	
143 619	**FI**	GW	*GW*	EX	55660	55685	
143 620	**GW**	P	*GW*	EX	55661	55686	
143 621	**GW**	P	*GW*	EX	55662	55687	
143 622	**AW**	P	*AW*	CF	55663	55688	
143 623	**AW**	P	*AW*	CF	55664	55689	
143 624	**AW**	P	*AW*	CF	55665	55690	
143 625	**AW**	P	*AW*	CF	55666	55691	

CLASS 144   PACER   ALEXANDER/BREL DERBY

DMS–DMSL or DMS–MS–DMSL. As Class 143, but underframes built by BREL.

Construction: Steel underframe, aluminium alloy body and roof. Alexander bus bodywork on four-wheeled underframes.
Engines: One Cummins LT10-R of 165 kW (225 hp) at 1950 rpm.
Couplers: BSI at outer ends, bar within unit.
Dimensions: 15.45/15.43 x 2.80 m.
Gangways: Within unit only.   **Wheel Arrangement:** 1-A + A-1.
Doors: Twin-leaf inward pivoting.   **Maximum Speed:** 75 mph.
Seating Layout: 2+2 high-back Richmond seating, mainly unidirectional.
Multiple Working: Within class and with Classes 142, 143, 150, 153, 155, 156, 158 and 159.

† Prototype demonstrator unit, refurbished by RVEL for Porterbrook as a trial, with new Fainsa seating and a universal access toilet to meet the 2020 accessibility regulations. Details are as follows:
DMS 55812: Lot No. 31015 BREL Derby 1986–87. –/43(+3). 27.2 t
DMSL 55835: Lot No. 31016 BREL Derby 1986–87. –/35 1TD 2W. 28.0 t.

Non-standard livery: 144 012 144evolution (blue & purple).

DMS. Lot No. 31015 BREL Derby 1986–87. –/45(+3) 1W 24.0 t.
MS. Lot No. 31037 BREL Derby 1987. –/58. 23.5 t.
DMSL. Lot No. 31016 BREL Derby 1986–87. –/41(+3) 1T. 24.5 t.

144 001	**NO**	P	*NO*	NL	55801	55824
144 002	**NO**	P	*NO*	NL	55802	55825
144 003	**NO**	P	*NO*	NL	55803	55826
144 004	**NO**	P	*NO*	NL	55804	55827
144 005	**NO**	P	*NO*	NL	55805	55828

▲ London Midland-liveried Parry People Mover 139 002 arrives at Stourbridge Junction with a shuttle from Stourbridge Town on 21/09/14. **Robert Pritchard**

▼ Arriva Trains-liveried 142 076 is seen at Tenby with the 11.50 Swansea–Pembroke Dock on 11/04/15. **Tom McAtee**

▲ Ex-works in Great Western Railway Green livery, 143 603 is seen with 153 373 at Newton Abbot on 22/07/17 with the 09.04 Paignton–Exmouth. **Tony Christie**

▼ Northern-liveried 144 002 arrives at Saxilby with the 09.19 Scunthorpe–Lincoln Central on 29/12/16. **Robert Pritchard**

▲ Arriva Trains Wales dark blue-liveried 150 240 stands at Blaenau Ffestiniog with the 15.03 to Llandudno Junction on 13/08/17.

Tony Christie

▲ In the new Northern livery, 150 137 is seen at York with the 08.47 to Leeds via Harrogate on 13/07/17. **Gavin Morrison**

▼ Greater Anglia-liveried 153 335 arrives at Great Yarmouth with the 07.36 from Norwich on 14/06/17. **Robert Pritchard**

▲ New Northern-liveried 155 346 stands at Leeds on 02/08/17.   **James Stokes**

▼ ScotRail Saltire-liveried 156 442 is seen near Breich with the 19.27 Edinburgh Waverley–Glasgow Central on 11/07/17.   **Robin Ralston**

▲ Great Western Railway Green-liveried 158 957 approaches Southampton Central with the 13.30 Cardiff Central–Portsmouth Harbour on 17/06/17. **Robert Pritchard**

▼ South West Trains-liveried 159 103 and 159 001 are seen near Wimbledon with the 06.19 Honiton–London Waterloo on 19/06/17. **Robert Pritchard**

▲ Chiltern Railways-liveried 165 020 leaves Kings Sutton with the 14.47 Banbury–London Marylebone on 22/05/17. **Brian Denton**

▼ Great Western Railway Green-liveried 166 217 calls at Ealing Broadway with the 10.48 Reading–London Paddington on 02/09/17. **Robert Pritchard**

▲ Chiltern Railways Mainline-liveried 168 112 and 168 321 pass Neasden with the 16.18 London Marylebone–Oxford on 05/07/16. **Robert Pritchard**

▼ Greater Anglia-liveried 170 270 arrives at March with the 13.58 Ipswich–Peterborough on 12/07/17. **Antony Guppy**

144 006	**NO**	P	*NO*	NL	55806		55829
144 007	**NO**	P	*NO*	NL	55807		55830
144 008	**NO**	P	*NO*	NL	55808		55831
144 009	**NO**	P	*NO*	NL	55809		55832
144 010	**NO**	P	*NO*	NL	55810		55833
144 011	**NO**	P	*NO*	NL	55811		55834
144 012 †	**0**	P	*NO*	NL	55812		55835
144 013	**NO**	P	*NO*	NL	55813		55836
144 014	**NO**	P	*NO*	NL	55814	55850	55837
144 015	**NO**	P	*NO*	NL	55815	55851	55838
144 016	**NO**	P	*NO*	NL	55816	55852	55839
144 017	**NO**	P	*NO*	NL	55817	55853	55840
144 018	**NO**	P	*NO*	NL	55818	55854	55841
144 019	**NO**	P	*NO*	NL	55819	55855	55842
144 020	**NO**	P	*NO*	NL	55820	55856	55843
144 021	**NO**	P	*NO*	NL	55821	55857	55844
144 022	**NO**	P	*NO*	NL	55822	55858	55845
144 023	**NO**	P	*NO*	NL	55823	55859	55846

Name: 144 001 THE PENISTONE LINE PARTNERSHIP

CLASS 150/0   SPRINTER   BREL YORK

DMSL–MS–DMS. Prototype Sprinter.

Construction: Steel.
Engines: One Cummins NT855R5 of 213 kW (285 hp) at 2100 rpm.
Bogies: BX8P (powered), BX8T (non-powered).
Couplers: BSI at outer end of driving vehicles, bar non-driving ends.
Dimensions: 19.93/19.92 x 2.73 m.
Gangways: Within unit only.   **Wheel Arrangement:** 2-B + 2-B + B-2.
Doors: Twin-leaf sliding.   **Maximum Speed:** 75 mph.
Seating Layout: 3+2 (mainly unidirectional).
Multiple Working: Within class and with Classes 142, 143, 144, 153, 155, 156, 158, 159, 170 and 172.

DMSL. Lot No. 30984 1984. –/72 1T. 35.4 t.
MS. Lot No. 30986 1984. –/92. 35.0 t.
DMS. Lot No. 30985 1984. –/69(+6). 34.7 t.

| 150 001 | **GW** | A | *GW* | PM | 55200 | 55400 | 55300 |
| 150 002 | **GW** | A | *GW* | PM | 55201 | 55401 | 55301 |

CLASS 150/1   SPRINTER   BREL YORK

DMSL–DMS.

Construction: Steel.
Engines: One Cummins NT855R5 of 213 kW (285 hp) at 2100 rpm.
Bogies: BP38 (powered), BT38 (non-powered).
Couplers: BSI.
Dimensions: 19.74 x 2.82 m.
Gangways: Within unit only.   **Wheel Arrangement:** 2-B (+ 2-B) + B-2.

Doors: Twin-leaf sliding.   **Maximum Speed:** 75 mph.
Seating Layout: 3+2 facing as built but Centro units were reseated with mainly unidirectional seating.
Multiple Working: Within class and with Classes 142, 143, 144, 153, 155, 156, 158, 159, 170 and 172.

c 3+2 Chapman seating.
* Northern units fitted with a new universal access toilet to meet the 2020 accessibility regulations. Full details awaited.

DMSL. Lot No. 31011 1985–86. –/72 1T (c –/59 1TD (except 52144 –/62 1TD), t –/71 1T, u –/71 1T). 38.3 t.
DMS. Lot No. 31012 1985–86. –/76 (c –/65, t –/73, u –/70(+6)). 38.1 t.

150 101	u	**FB**	A	*GW*	PM	52101	57101
150 102	u	**FB**	A	*GW*	PM	52102	57102
150 103	u	**NO**	A	*NO*	NH	52103	57103
150 104	u	**FB**	A	*GW*	PM	52104	57104
150 105	u	**LM**	A	*LM*	TS	52105	57105
150 106	u	**FB**	A	*GW*	PM	52106	57106
150 107	u	**LM**	A	*LM*	TS	52107	57107
150 108	u	**FB**	A	*GW*	PM	52108	57108
150 109	u	**LM**	A	*LM*	TS	52109	57109
150 110	u	**NO**	A	*NO*	NH	52110	57110
150 111	u	**NO**	A	*NO*	NH	52111	57111
150 112	u	**NO**	A	*NO*	NH	52112	57112
150 113	u	**NO**	A	*NO*	NH	52113	57113
150 114	u	**NO**	A	*NO*	NH	52114	57114
150 115	u	**NO**	A	*NO*	NH	52115	57115
150 116	u	**NO**	A	*NO*	NH	52116	57116
150 117	u	**NO**	A	*NO*	NH	52117	57117
150 118	u	**NO**	A	*NO*	NH	52118	57118
150 119	u	**NO**	A	*NO*	NH	52119	57119
150 120	t	**FB**	A	*GW*	EX	52120	57120
150 121	u	**FB**	A	*GW*	EX	52121	57121
150 122	u	**FB**	A	*GW*	PM	52122	57122
150 123	t	**FB**	A	*GW*	EX	52123	57123
150 124	u	**FB**	A	*GW*	PM	52124	57124
150 127	u	**FB**	A	*GW*	EX	52127	57127
150 128	t	**FB**	A	*GW*	EX	52128	57128
150 129	t	**FB**	A	*GW*	EX	52129	57129
150 130	t	**FB**	A	*GW*	EX	52130	57130
150 131	t	**FB**	A	*GW*	EX	52131	57131
150 132	u	**NO**	A	*NO*	NH	52132	57132
150 133	c	**NO**	A	*NO*	NH	52133	57133
150 134	c*	**NR**	A	*NO*	NH	52134	57134
150 135	c	**NO**	A	*NO*	NH	52135	57135
150 136	c*	**NR**	A	*NO*	NH	52136	57136
150 137	c*	**NR**	A	*NO*	NH	52137	57137
150 138	c	**NO**	A	*NO*	NH	52138	57138
150 139	c	**NO**	A	*NO*	NH	52139	57139
150 140	c*	**NR**	A	*NO*	NH	52140	57140

150 141	c	**NO**	A	*NO*	NH	52141	57141
150 142	c*	**NR**	A	*NO*	NH	52142	57142
150 143	c	**NO**	A	*NO*	NH	52143	57143
150 144	c	**NO**	A	*NO*	NH	52144	57144
150 145	c	**NO**	A	*NO*	NH	52145	57145
150 146	c	**NO**	A	*NO*	NH	52146	57146
150 147	c	**NO**	A	*NO*	NH	52147	57147
150 148	c	**NO**	A	*NO*	NH	52148	57148
150 149	c	**NO**	A	*NO*	NH	52149	57149
150 150	c	**NO**	A	*NO*	NH	52150	57150

Names:

| 150 129 | Devon & Cornwall RAIL PARTNERSHIP |
| 150 130 | Severnside Community Rail Partnership |

CLASS 150/2   SPRINTER   BREL YORK

DMSL–DMS.

Construction: Steel.
Engines: One Cummins NT855R5 of 213 kW (285 hp) at 2100 rpm.
Bogies: BP38 (powered), BT38 (non-powered).
Couplers: BSI.
Dimensions: 19.74 x 2.82 m.
Gangways: Throughout.
Doors: Twin-leaf sliding.
Wheel Arrangement: 2-B + B-2.
Maximum Speed: 75 mph.
Seating Layout: 3+2 mainly unidirectional seating as built, but most units have been refurbished with new 2+2 seating.
Multiple Working: Within class and with Classes 142, 143, 144, 153, 155, 156, 158, 159, 170 and 172.

c   Former First North Western units with 3+2 Chapman seating.
p   Refurbished Arriva Trains Wales units with 2+2 Primarius seating.
v   Units refurbished for Valley Lines with 2+2 Chapman seating.
w   Units refurbished for First Great Western with 2+2 Chapman seating.
*   Refurbished units for Great Western Railway with 2+2 Chapman seating and a new universal access toilet to meet the 2020 accessibility regulations.
†   Refurbished units for Northern with a new universal access toilet to meet the 2020 accessibility regulations. 3+2 Chapman seating.
§   Refurbished units for Northern with new universal access toilet to meet the 2020 accessibility regulations. Orignal Ashbourne seating. Full detais awaited.

Northern promotional vinyls:

150 203/205/207/211/215/218/222/223/225/228/268–271/273/274/277
Welcome to Yorkshire.
150 272 R&B Festival week, Colne.

DMSL. Lot No. 31017 1986–87. * –/50(+4) 1TD 2W, † –/58(+3) 1TD 2W, c –/62 1TD, p –/60(+4) 1T, s –/68 1T 1W, u –/71 1T), v –/60(+8) 1T, w –/60(+8) 1T. 37.5 t (* 35.8 t, t 38.1 t).
DMS. Lot No. 31018 1986–87. * –/58(10), † –/70(+6), c –/70, p –/56(+10) 1W, s –/71(+3), †u –/70(+6), v –/56(+15) 2W, w –/56(+17) 2W. 36.5 t.

150 201	c	**NO**	A	*NO*	NH	52201	57201
150 202	u	**GW**	A	*GW*	PM	52202	57202
150 203	†	**NR**	A	*NO*	NH	52203	57203
150 204	†	**NO**	A	*NO*	NH	52204	57204
150 205	†	**NR**	A	*NO*	NH	52205	57205
150 206	†	**NO**	A	*NO*	NH	52206	57206
150 207	c	**NO**	A	*NO*	NH	52207	57207
150 208	p	**AW**	P	*AW*	CF	52208	57208
150 210	u	**NO**	A	*NO*	NH	52210	57210
150 211	†	**NO**	A	*NO*	NH	52211	57211
150 213	p	**AW**	P	*AW*	CF	52213	57213
150 214	u	**NO**	A	*NO*	NH	52214	57214
150 215	c	**NO**	A	*NO*	NH	52215	57215
150 216	u	**GW**	A	*GW*	PM	52216	57216
150 217	p	**AW**	P	*AW*	CF	52217	57217
150 218	c	**NO**	A	*NO*	NH	52218	57218
150 219	*	**FB**	P	*GW*	PM	52219	57219
150 220	†	**NR**	A	*NO*	NH	52220	57220
150 221	w	**FI**	P	*GW*	PM	52221	57221
150 222	†	**NO**	A	*NO*	NH	52222	57222
150 223	†	**NO**	A	*NO*	NH	52223	57223
150 224	c	**NO**	A	*NO*	NH	52224	57224
150 225	c	**NO**	A	*NO*	NH	52225	57225
150 226	u	**NO**	A	*NO*	NH	52226	57226
150 227	p	**AW**	P	*AW*	CF	52227	57227
150 228	s	**NO**	P	*NO*	NH	52228	57228
150 229	p	**AW**	P	*AW*	CF	52229	57229
150 230	w	**AW**	P	*AW*	CF	52230	57230
150 231	p	**AW**	P	*AW*	CF	52231	57231
150 232	*	**GW**	P	*GW*	PM	52232	57232
150 233	*	**GW**	P	*GW*	PM	52233	57233
150 234	*	**GW**	P	*GW*	PM	52234	57234
150 235	p	**AW**	P	*AW*	CF	52235	57235
150 236	w	**AW**	P	*AW*	CF	52236	57236
150 237	p	**AW**	P	*AW*	CF	52237	57237
150 238	*	**FB**	P	*GW*	PM	52238	57238
150 239	*	**GW**	P	*GW*	PM	52239	57239
150 240	w	**AW**	P	*AW*	CF	52240	57240
150 241	w	**AW**	P	*AW*	CF	52241	57241
150 242	w	**AW**	P	*AW*	CF	52242	57242
150 243	*	**GW**	P	*GW*	PM	52243	57243
150 244	*	**GW**	P	*GW*	PM	52244	57244
150 245	p	**AV**	P	*AW*	CF	52245	57245
150 246	*	**GW**	P	*GW*	PM	52246	57246
150 247	*	**GW**	P	*GW*	PM	52247	57247
150 248	*	**GW**	P	*GW*	PM	52248	57248
150 249	*	**GW**	P	*GW*	PM	52249	57249
150 250	p	**AW**	P	*AW*	CF	52250	57250
150 251	w	**AW**	P	*AW*	CF	52251	57251
150 252	p	**AW**	P	*AW*	CF	52252	57252
150 253	w	**AW**	P	*AW*	CF	52253	57253

150 254	w	**AW**	P	*AW*	CF	52254	57254
150 255	p	**AW**	P	*AW*	CF	52255	57255
150 256	p	**AW**	P	*AW*	CF	52256	57256
150 257	p	**AW**	P	*AW*	CF	52257	57257
150 258	p	**AV**	P	*AW*	CF	52258	57258
150 259	p	**AV**	P	*AW*	CF	52259	57259
150 260	p	**AW**	P	*AW*	CF	52260	57260
150 261	*	**GW**	P	*GW*	PM	52261	57261
150 262	p	**AW**	P	*AW*	CF	52262	57262
150 263	*	**GW**	P	*GW*	PM	52263	57263
150 264	p	**AV**	P	*AW*	CF	52264	57264
150 265	w	**FI**	P	*GW*	PM	52265	57265
150 266	*	**GW**	P	*GW*	PM	52266	57266
150 267	v	**AW**	P	*AW*	CF	52267	57267
150 268	s	**NO**	P	*NO*	NH	52268	57268
150 269	s	**NO**	P	*NO*	NH	52269	57269
150 270	s	**NO**	P	*NO*	NH	52270	57270
150 271	s	**NO**	P	*NO*	NH	52271	57271
150 272	s	**NO**	P	*NO*	NH	52272	57272
150 273	s	**NO**	P	*NO*	NH	52273	57273
150 274	§	**NR**	P	*NO*	NH	52274	57274
150 275	§	**NR**	P	*NO*	NH	52275	57275
150 276	§	**NR**	P	*NO*	NH	52276	57276
150 277	s	**NO**	P	*NO*	NH	52277	57277
150 278	v	**AW**	P	*AW*	CF	52278	57278
150 279	v	**AW**	P	*AW*	CF	52279	57279
150 280	v	**AW**	P	*AW*	CF	52280	57280
150 281	v	**AW**	P	*AW*	CF	52281	57281
150 282	v	**AW**	P	*AW*	CF	52282	57282
150 283	p	**AW**	P	*AW*	CF	52283	57283
150 284	p	**AV**	P	*AW*	CF	52284	57284
150 285	p	**AV**	P	*AW*	CF	52285	57285

CLASS 150/9   SPRINTER   BREL YORK

3-car Great Western Railway hybrids formed of a Class 150/1 with a 150/2 centre vehicle. DMSL–DMS–DMS. For details see Class 150/1 or Class 150/2.

| 150 925 | u | **FB** | A | *GW* | PM | 52125 | 57209 | 57125 |
| 150 926 | u | **FB** | A | *GW* | PM | 52126 | 57212 | 57126 |

Name:

150 925   THE HEART OF WESSEX LINE

CLASS 153   SUPER SPRINTER   LEYLAND BUS

DMSL. Converted by Hunslet-Barclay, Kilmarnock from Class 155 2-car units.

Construction: Steel underframe, rivetted steel body and roof. Built from Leyland National bus parts on Leyland Bus bogied underframes.
Engine: One Cummins NT855R5 of 213 kW (285 hp) at 2100 rpm.
Bogies: One P3-10 (powered) and one BT38 (non-powered).
Couplers: BSI.
Dimensions: 23.21 x 2.70 m.
Gangways: Throughout.   **Wheel Arrangement:** 2-B.
Doors: Single-leaf sliding plug.   **Maximum Speed:** 75 mph.
Seating Layout: 2+2 facing/unidirectional.
Multiple Working: Within class and with Classes 142, 143, 144, 150, 155, 156, 158, 159, 170 and 172.

Cars numbered in the 573xx series were renumbered by adding 50 to their original number so that the last two digits correspond with the set number.

c   Chapman seating.
d   Richmond seating.

Non-standard and Advertising liveries:

153 305   Light grey.
153 325   Citizens Rail (pink).
153 333   Visit South Devon by train (dark blue).

52301–52335. DMSL. Lot No. 31026 1987–88. Converted under Lot No. 31115 1991–92. –/72(+3) 1T 1W. (s –/72 1T 1W, t –/72(+2) 1T 1W). 41.2 t.
57301–57335. DMSL. Lot No. 31027 1987–88. Converted under Lot No. 31115 1991–92. –/72(+3) 1T 1W (s –/72 1T 1W). 41.2 t.

153 301	d	**NO**	A	*NO*	NL	52301	
153 302	c	**EM**	A	*EM*	NM	52302	
153 303	c	**AW**	A	*AW*	CF	52303	
153 304	ds	**NO**	A	*NO*	NL	52304	
153 305	d	**O**	A	*GW*	EX	52305	
153 306	cr	**GA**	P	*GA*	NC	52306	
153 307	d	**NO**	A	*NO*	NL	52307	
153 308	c	**EM**	A	*EM*	NM	52308	
153 309	cr	**GA**	P	*GA*	NC	52309	GERARD FIENNES
153 310	c	**EM**	P	*EM*	NM	52310	
153 311	c	**EM**	P	*EM*	NM	52311	
153 312	s	**AW**	A	*AW*	CF	52312	
153 313	cs	**EM**	P	*EM*	NM	52313	
153 314	cr	**GA**	P	*GA*	NC	52314	
153 315	ds	**NO**	A	*NO*	NL	52315	
153 316	c	**NO**	P	*NO*	NL	52316	John "Longitude" Harrison Inventor of the Marine Chronometer
153 317	ds	**NO**	A	*NO*	NL	52317	
153 318	d	**GW**	A	*GW*	EX	52318	
153 319	c	**EM**	A	*EM*	NM	52319	
153 320	c	**AV**	P	*AW*	CF	52320	

153 321	ct	**EM**	P	*EM*	NM	52321	
153 322	cr	**GA**	P	*GA*	NC	52322	BENJAMIN BRITTEN
153 323	c	**AV**	P	*AW*	CF	52323	
153 324	c	**NO**	P	*NO*	NL	52324	
153 325	c	**AL**	P	*GW*	EX	52325	
153 326	c	**EM**	P	*EM*	NM	52326	
153 327	c	**AW**	A	*AW*	CF	52327	
153 328	ds	**NO**	A	*NO*	NL	52328	
153 329	c	**FB**	P	*GW*	EX	52329	
153 330	cs	**NO**	P	*NO*	NL	52330	
153 331	d	**NO**	A	*NO*	NL	52331	
153 332	c	**NO**	P	*NO*	NL	52332	
153 333	cs	**AL**	P	*GW*	EX	52333	
153 334	ct	**LM**	P	*LM*	TS	52334	
153 335	cr	**GA**	P	*GA*	NC	52335	MICHAEL PALIN
153 351	d	**NO**	A	*NO*	NL	57351	
153 352	ds	**NO**	A	*NO*	NL	57352	
153 353	c	**AW**	A	*AW*	CF	57353	
153 354	c	**LM**	P	*LM*	TS	57354	
153 355	c	**EM**	A	*EM*	NM	57355	
153 356	c	**LM**	P	*LM*	TS	57356	
153 357	c	**EM**	A	*EM*	NM	57357	
153 358	c	**NO**	P	*NO*	NL	57358	
153 359	c	**NO**	P	*NO*	NL	57359	
153 360	c	**NO**	P	*NO*	NL	57360	
153 361	cs	**FB**	P	*GW*	EX	57361	
153 362	cs	**AW**	A	*AW*	CF	57362	
153 363	cs	**NO**	P	*NO*	NL	57363	
153 364	c	**LM**	P	*LM*	TS	57364	
153 365	c	**LM**	P	*LM*	TS	57365	
153 366	c	**LM**	P	*LM*	TS	57366	
153 367	cs	**AV**	P	*AW*	CF	57367	
153 368	d	**GW**	A	*GW*	EX	57368	
153 369	c	**FB**	P	*GW*	EX	57369	
153 370	d	**GW**	A	*GW*	EX	57370	
153 371	c	**LM**	P	*LM*	TS	57371	
153 372	d	**GW**	A	*GW*	EX	57372	
153 373	d	**GW**	A	*GW*	EX	57373	
153 374	c	**EM**	A	*EM*	NM	57374	
153 375	c	**LM**	P	*LM*	TS	57375	
153 376	c	**EM**	P	*EM*	NM	57376	X24-EXPEDITIOUS
153 377	d	**GW**	A	*GW*	EX	57377	
153 378	d	**NO**	A	*NO*	NL	57378	
153 379	c	**EM**	P	*EM*	NM	57379	
153 380	d	**GW**	A	*GW*	EX	57380	
153 381	c	**EM**	P	*EM*	NM	57381	
153 382	d	**GW**	A	*GW*	EX	57382	
153 383	c	**EM**	P	*EM*	NM	57383	Ecclesbourne Valley Railway 150 Years
153 384	c	**EM**	P	*EM*	NM	57384	
153 385	c	**EM**	P	*EM*	NM	57385	

CLASS 155   SUPER SPRINTER   LEYLAND BUS

DMSL–DMS.

Construction: Steel underframe, rivetted steel body and roof. Built from Leyland National bus parts on Leyland Bus bogied underframes.
Engines: One Cummins NT855R5 of 213 kW (285 hp) at 2100 rpm.
Bogies: One P3-10 (powered) and one BT38 (non-powered).
Couplers: BSI.
Dimensions: 23.21 x 2.70 m.

Gangways: Throughout.	**Wheel Arrangement:** 2-B + B-2.
Doors: Single-leaf sliding plug.	**Maximum Speed:** 75 mph.

Seating Layout: 2+2 facing/unidirectional Chapman seating.
Multiple Working: Within class and with Classes 142, 143, 144, 150, 153, 156, 158, 159, 170 and 172.

Northern promotional vinyls:

155 341–344/347 Leeds–Bradford–Manchester route (the "Calder Valley").

DMSL. Lot No. 31057 1988. –/76 1TD 1W. 39.0 t.
DMS. Lot No. 31058 1988. –/80. 38.6 t.

155 341	**NO**	P	*NO*	NL	52341	57341
155 342	**NO**	P	*NO*	NL	52342	57342
155 343	**NO**	P	*NO*	NL	52343	57343
155 344	**NO**	P	*NO*	NL	52344	57344
155 345	**NR**	P	*NO*	NL	52345	57345
155 346	**NR**	P	*NO*	NL	52346	57346
155 347	**NO**	P	*NO*	NL	52347	57347

CLASS 156   SUPER SPRINTER   METRO-CAMMELL

DMSL–DMS.

Construction: Steel.
Engines: One Cummins NT855R5 of 213 kW (285 hp) at 2100 rpm.
Bogies: One P3-10 (powered) and one BT38 (non-powered).
Couplers: BSI.
Dimensions: 23.03 x 2.73 m.

Gangways: Throughout.	**Wheel Arrangement:** 2-B + B-2.
Doors: Single-leaf sliding.	**Maximum Speed:** 75 mph.

Seating Layout: 2+2 facing/unidirectional.
Multiple Working: Within class and with Classes 142, 143, 144, 150, 153, 155, 158, 159, 170 and 172.

† Greater Anglia units fitted with a new universal access toilet to meet the 2020 accessibility regulations.
* Angel-owned Northern units fitted with a new universal access toilet to meet the 2020 accessibility regulations.
b Refurbished as a demonstrator unit by Brodies, Kilmarnock with a new universal access toilet to meet the 2020 accessibility regulations. Full details awaited.

c Chapman seating.
d Richmond seating.
v ScotRail units fitted with a new universal access toilet to meet the 2020 accessibility regulations (but still with original seating).
w Fully refurbished ScotRail units with new Fainsa seating and universal access toilet. Full details awaited.

Northern promotional vinyls:

156441 Manchester and Liverpool
156461 Ravenglass & Eskdale Railway
156464 Lancashire DalesRail

DMSL. Lot No. 31028 1988–89. –/74 1TD 1W († –/62 1TD 2W, * –/64(+2) 1TD 2W, c –/70, s –/72, t –/68 1W, u –/68, v –/64(+3) 1TD 2W). 38.6 t.
DMS. Lot No. 31029 1987–89. –/76 († –/74, *–/76, d –/78 , t & u –/72, v –/76). 36.1 t.

156 401	cs	**EM**	P	*EM*	DY	52401	57401
156 402	†cr	**GA**	P	*GA*	NC	52402	57402
156 403	cs	**EM**	P	*EM*	DY	52403	57403
156 404	cs	**EM**	P	*EM*	DY	52404	57404
156 405	cs	**EM**	P	*EM*	DY	52405	57405
156 406	cs	**EM**	P	*EM*	DY	52406	57406
156 407	†cr	**GA**	P	*GA*	NC	52407	57407
156 408	cs	**EM**	P	*EM*	DY	52408	57408
156 409	†cr	**GA**	P	*GA*	NC	52409	57409
156 410	cs	**EM**	P	*EM*	DY	52410	57410
156 411	cs	**EM**	P	*EM*	DY	52411	57411
156 412	†cr	**GA**	P	*GA*	NC	52412	57412
156 413	cs	**EM**	P	*EM*	DY	52413	57413
156 414	cs	**EM**	P	*EM*	DY	52414	57414
156 415	cs	**EM**	P	*EM*	DY	52415	57415
156 416	†cr	**GA**	P	*GA*	NC	52416	57416
156 417	†cr	**GA**	P	*GA*	NC	52417	57417
156 418	†cr	**GA**	P	*GA*	NC	52418	57418
156 419	†cr	**GA**	P	*GA*	NC	52419	57419
156 420	c	**NO**	P	*NO*	AN	52420	57420
156 421	c	**NO**	P	*NO*	AN	52421	57421
156 422	†cr	**GA**	P	*GA*	NC	52422	57422
156 423	c	**NO**	P	*NO*	AN	52423	57423
156 424	c	**NO**	P	*NO*	AN	52424	57424
156 425	c	**NO**	P	*NO*	AN	52425	57425
156 426	c	**NO**	P	*NO*	AN	52426	57426
156 427	c	**NO**	P	*NO*	AN	52427	57427
156 428	c	**NO**	P	*NO*	AN	52428	57428
156 429	c	**NO**	P	*NO*	AN	52429	57429
156 430	t	**SR**	A	*SR*	CK	52430	57430
156 431	t	**SR**	A	*SR*	CK	52431	57431
156 432	v	**SR**	A	*SR*	CK	52432	57432
156 433	t	**SR**	A	*SR*	CK	52433	57433
156 434	t	**SR**	A	*SR*	CK	52434	57434
156 435	t	**SR**	A	*SR*	CK	52435	57435
156 436	†	**SR**	A	*SR*	CK	52436	57436

156 437	t	**SR**	A	*SR*	CK	52437	57437
156 438	*d	**NR**	A	*NO*	HT	52438	57438
156 439	t	**SR**	A	*SR*	CK	52439	57439
156 440	c	**NO**	P	*NO*	AN	52440	57440
156 441	c	**NO**	P	*NO*	AN	52441	57441
156 442	t	**SR**	A	*SR*	CK	52442	57442
156 443	*d	**NR**	A	*NO*	HT	52443	57443
156 444	*d	**NR**	A	*NO*	HT	52444	57444
156 445	ru	**SR**	A	*SR*	CK	52445	57445
156 446	t	**FS**	A	*SR*	CK	52446	57446
156 447	ru	**FS**	A	*SR*	CK	52447	57447
156 448	*d	**NO**	A	*NO*	HT	52448	57448
156 449	u	**FS**	A	*SR*	CK	52449	57449
156 450	ru	**FS**	A	*SR*	CK	52450	57450
156 451	*d	**NR**	A	*NO*	HT	52451	57451
156 452	c	**NO**	P	*NO*	AN	52452	57452
156 453	ru	**FS**	A	*SR*	CK	52453	57453
156 454	*d	**NR**	A	*NO*	HT	52454	57454
156 455	c	**NO**	P	*NO*	AN	52455	57455
156 456	rt	**FS**	A	*SR*	CK	52456	57456
156 457	rt	**FS**	A	*SR*	CK	52457	57457
156 458	rt	**FS**	A	*SR*	CK	52458	57458
156 459	c	**NO**	P	*NO*	AN	52459	57459
156 460	c	**NO**	P	*NO*	AN	52460	57460
156 461	c	**NO**	P	*NO*	AN	52461	57461
156 462		**FS**	A	*SR*	CK	52462	57462
156 463	*d	**NR**	A	*NO*	HT	52463	57463
156 464	c	**NO**	P	*NO*	AN	52464	57464
156 465	ru	**FS**	A	*SR*	CK	52465	57465
156 466	c	**NO**	P	*NO*	AN	52466	57466
156 467	rv	**SR**	A	*SR*	CK	52467	57467
156 468	d	**NO**	A	*NO*	AN	52468	57468
156 469	d	**NO**	A	*NO*	HT	52469	57469
156 470	c	**EM**	A	*EM*	DY	52470	57470
156 471	*d	**NR**	A	*NO*	AN	52471	57471
156 472	*d	**NR**	A	*NO*	AN	52472	57472
156 473	c	**EM**	A	*EM*	DY	52473	57473
156 474	rt	**FS**	A	*SR*	CK	52474	57474
156 475	*d	**NO**	A	*NO*	HT	52475	57475
156 476	rt	**FS**	A	*SR*	CK	52476	57476
156 477	t	**FS**	A	*SR*	CK	52477	57477
156 478	b	**SR**	BR	*SR*	CK	52478	57478
156 479	*d	**NR**	A	*NO*	HT	52479	57479
156 480	*d	**NO**	A	*NO*	HT	52480	57480
156 481	*d	**NR**	A	*NO*	HT	52481	57481
156 482	*d	**NR**	A	*NO*	AN	52482	57482
156 483	*d	**NO**	A	*NO*	AN	52483	57483
156 484	*d	**NR**	A	*NO*	HT	52484	57484
156 485	ru	**FS**	A	*SR*	CK	52485	57485
156 486	*d	**NR**	A	*NO*	AN	52486	57486
156 487	*d	**NO**	A	*NO*	AN	52487	57487

156 488	*d	**NR**	A	*NO*	AN	52488	57488
156 489	*d	**NR**	A	*NO*	AN	52489	57489
156 490	d	**NO**	A	*NO*	HT	52490	57490
156 491	*d	**NO**	A	*NO*	AN	52491	57491
156 492	rs	**SR**	A	*SR*	CK	52492	57492
156 493	rt	**FS**	A	*SR*	CK	52493	57493
156 494	u	**SR**	A	*SR*	CK	52494	57494
156 495	v	**SR**	A	*SR*	CK	52495	57495
156 496	ru	**FS**	A	*SR*	CK	52496	57496
156 497	c	**EM**	A	*EM*	DY	52497	57497
156 498	c	**EM**	A	*EM*	DY	52498	57498
156 499	rt	**SR**	A	*SR*	CK	52499	57499
156 500	ru	**SR**	A	*SR*	CK	52500	57500
156 501	v	**SR**	A	*SR*	CK	52501	57501
156 502	v	**SR**	A	*SR*	CK	52502	57502
156 503	v	**SR**	A	*SR*	CK	52503	57503
156 504		**SR**	A	*SR*	CK	52504	57504
156 505	v	**SR**	A	*SR*	CK	52505	57505
156 506	v	**SR**	A	*SR*	CK	52506	57506
156 507	v	**SR**	A	*SR*	CK	52507	57507
156 508	v	**SR**	A	*SR*	CK	52508	57508
156 509	v	**SR**	A	*SR*	CK	52509	57509
156 510	w	**SR**	A	*SR*	CK	52510	57510
156 511	v	**SR**	A	*SR*	CK	52511	57511
156 512		**SR**	A	*SR*	CK	52512	57512
156 513	v	**SR**	A	*SR*	CK	52513	57513
156 514	v	**SR**	A	*SR*	CK	52514	57514

Names:

156 416	Saint Edmund
156 418	ESTA 1965–2015
156 420	LA' AL RATTY Ravenglass & Eskdale Railway
156 440	George Bradshaw
156 441	William Huskisson MP
156 448	Bram Stoker Creator of Dracula
156 459	Benny Rothman – The Manchester Rambler
156 460	Driver John Axon G.C.
156 464	Lancashire DalesRail
156 466	Gracie Fields
156 469	The Royal Northumberland Fusiliers (The Fighting Fifth)
156 490	Captain James Cook Master Mariner

CLASS 158/0   BREL

DMSL(B)–DMSL(A) or DMCL–DMSL or DMSL–MSL–DMSL.

Construction: Welded aluminium.
Engines: 158 701–813/158 880–890/158 950–961: One Cummins NTA855R1 of 260 kW (350 hp) at 2100 rpm.
158 815–862: One Perkins 2006-TWH of 260 kW (350 hp) at 2100 rpm.
158 863–872: One Cummins NTA855R3 of 300 kW (400 hp) at 1900 rpm.

Bogies: One BREL P4 (powered) and one BREL T4 (non-powered) per car.
Couplers: BSI.   **Dimensions:** 22.57 x 2.70 m.
Gangways: Throughout.   **Wheel Arrangement:** 2-B + B-2.
Doors: Twin-leaf swing plug.   **Maximum Speed:** 90 mph.
Seating Layout: 2+2 facing/unidirectional in all Standard and First Class
except 2+1 in South West Trains First Class.
Multiple Working: Within class and with Classes 142, 143, 144, 150, 153,
155, 156, 159, 170 and 172.

ScotRail 158s 158 701–736/738–741 are "fitted" for RETB. When a unit
arrives at Inverness the cab display unit is clipped on and plugged in.

Arriva Trains Wales units have ERTMS plugged in at Shrewsbury for
working the Cambrian Lines.

*	Refurbished ScotRail units fitted with Grammer seating, additional luggage racks and cycle stowage areas. ScotRail units 158 726–736/738–741 are fitted with Richmond seating.
†	Refurbished East Midlands Trains units with Grammer seating.
§	Northern 3-car units with modifications for operation beyond 2020.
c	Chapman seating.
s	Refurbished Arriva Trains Wales units with Grammer seating.
t	East Midlands Trains units with modifications for operation beyond 2020.
u	Refurbished former South West Trains units with Class 159-style interiors, including First Class seating.
v	Northern units with modifications for operation beyond 2020.
w	Refurbished Great Western Railway units. Units 158 745–749/751/762/767 (now formed into 3-car sets) have Richmond seating.

Advertising livery:

158 798   Springboard Opportunity Group (light blue).

Northern promotional vinyls:

158 784 PTEG: 40 years.
158 787, 158 792–796: Sheffield–Leeds fast service.
158 790/842/843/844/845/848/849/850/853/855: We Are Northern.
158 860: Keighley & Brontë Country.
158 901–910: Leeds–Bradford–Manchester route (the "Calder Valley").

DMSL(B). Lot No. 31051 BREL Derby 1989–92. –/68 1TD 1W. († –/72 1TD 1W,
§, v –/64 1TD 1W, c, w –/66 1TD 1W, s –/64(+4) 1TD 2W, t – 68(+3) 1TD 2W,
z –/62 1TD 2W). 38.5 t.
MSL. Lot No. 31050 BREL Derby 1991. –/66(+3) 1T. 38.5 t.
DMSL(A). Lot No. 31052 BREL Derby 1989–92. –/70 1T († –/74, c, w –/68 1T,
* –/64(+2) 1T plus cycle stowage area). 38.5 t.

The above details refer to the "as built" condition. The following DMSL(B)
have now been converted to DMCL as follows:
52701–736/738–741 (ScotRail). 15/53 1TD 1W (* refurbished sets –/60(+6)
1TD 1W plus cycle stowage area).
52786/789 (Former South West Trains units). 13/44 1TD 1W.

158 701   *   **SR**   P   *SR*   IS   52701 57701

158 702	*	**SR**	P	*SR*	IS	52702	57702	
158 703	*	**SR**	P	*SR*	IS	52703	57703	
158 704	*	**SR**	P	*SR*	IS	52704	57704	
158 705	*	**SR**	P	*SR*	IS	52705	57705	
158 706	*	**SR**	P	*SR*	IS	52706	57706	
158 707	*	**SR**	P	*SR*	IS	52707	57707	
158 708	*	**SR**	P	*SR*	IS	52708	57708	
158 709	*	**SR**	P	*SR*	IS	52709	57709	
158 710	*	**SR**	P	*SR*	IS	52710	57710	
158 711	*	**SR**	P	*SR*	IS	52711	57711	
158 712	*	**SR**	P	*SR*	IS	52712	57712	
158 713	*	**SR**	P	*SR*	IS	52713	57713	
158 714	*	**SR**	P	*SR*	IS	52714	57714	
158 715	*	**SR**	P	*SR*	IS	52715	57715	
158 716	*	**SR**	P	*SR*	IS	52716	57716	
158 717	*	**SR**	P	*SR*	IS	52717	57717	
158 718	*	**SR**	P	*SR*	IS	52718	57718	
158 719	*	**SR**	P	*SR*	IS	52719	57719	
158 720	*	**SR**	P	*SR*	IS	52720	57720	
158 721	*	**SR**	P	*SR*	IS	52721	57721	
158 722	*	**SR**	P	*SR*	IS	52722	57722	
158 723	*	**SR**	P	*SR*	IS	52723	57723	
158 724	*	**SR**	P	*SR*	IS	52724	57724	
158 725	*	**SR**	P	*SR*	IS	52725	57725	
158 726		**FS**	P	*SR*	HA	52726	57726	
158 727		**FS**	P	*SR*	HA	52727	57727	
158 728		**FS**	P	*SR*	HA	52728	57728	
158 729		**FS**	P	*SR*	HA	52729	57729	
158 730		**FS**	P	*SR*	HA	52730	57730	
158 731		**SR**	P	*SR*	HA	52731	57731	
158 732		**FS**	P	*SR*	HA	52732	57732	
158 733		**FS**	P	*SR*	HA	52733	57733	
158 734		**FS**	P	*SR*	HA	52734	57734	
158 735		**FS**	P	*SR*	HA	52735	57735	
158 736		**FS**	P	*SR*	HA	52736	57736	
158 738		**FS**	P	*SR*	HA	52738	57738	
158 739		**FS**	P	*SR*	HA	52739	57739	
158 740		**FS**	P	*SR*	HA	52740	57740	
158 741		**FS**	P	*SR*	HA	52741	57741	
158 752	§	**NR**	P	*NO*	NL	52752	58716	57752
158 753	§	**NR**	P	*NO*	NL	52753	58710	57753
158 754	§	**NR**	P	*NO*	NL	52754	58708	57754
158 755	§	**NR**	P	*NO*	NL	52755	58702	57755
158 756	§	**NR**	P	*NO*	NL	52756	58712	57756
158 757		**NO**	P	*NO*	NL	52757	58706	57757
158 758		**NO**	P	*NO*	NL	52758	58714	57758
158 759		**NO**	P	*NO*	NL	52759	58713	57759
158 763	w	**FI**	P	*GW*	PM	52763	57763	
158 766	w	**GW**	P	*GW*	PM	52766	57766	
158 770	†	**ST**	P	*EM*	NM	52770	57770	
158 773	†	**ST**	P	*EM*	NM	52773	57773	

158 774	†	**ST**	P	*EM*	NM	52774	57774	
158 777	†	**ST**	P	*EM*	NM	52777	57777	
158 780	††	**ST**	A	*EM*	NM	52780	57780	
158 782		**SR**	A	*SR*	HA	52782	57782	
158 783	††	**ST**	A	*EM*	NM	52783	57783	
158 784		**NO**	A	*NO*	NL	52784	57784	
158 785	††	**ST**	A	*EM*	NM	52785	57785	
158 786	u	**SR**	A	*SR*	HA	52786	57786	
158 787		**NO**	A	*NO*	NL	52787	57787	
158 788	†	**ST**	A	*EM*	NM	52788	57788	
158 789	u	**SR**	A	*SR*	HA	52789	57789	
158 790		**NO**	A	*NO*	NL	52790	57790	
158 791		**NO**	A	*NO*	NL	52791	57791	
158 792		**NO**	A	*NO*	NL	52792	57792	
158 793		**NO**	A	*NO*	NL	52793	57793	
158 794		**NO**	A	*NO*	NL	52794	57794	
158 795		**NO**	A	*NO*	NL	52795	57795	
158 796		**NO**	A	*NO*	NL	52796	57796	
158 797		**NO**	A	*NO*	NL	52797	57797	
158 798	w	**AL**	P	*GW*	PM	52798	58715	57798
158 799	†	**ST**	P	*EM*	NM	52799	57799	
158 806	†	**ST**	P	*EM*	NM	52806	57806	
158 810	†	**ST**	P	*EM*	NM	52810	57810	
158 812	†	**ST**	P	*EM*	NM	52812	57812	
158 813	†	**ST**	P	*EM*	NM	52813	57813	
158 815	c	**NO**	A	*NO*	NL	52815	57815	
158 816	c	**NO**	A	*NO*	NL	52816	57816	
158 817	c	**NO**	A	*NO*	NL	52817	57817	
158 818	es	**AW**	A	*AW*	MN	52818	57818	
158 819	es	**AW**	A	*AW*	MN	52819	57819	
158 820	es	**AW**	A	*AW*	MN	52820	57820	
158 821	es	**AW**	A	*AW*	MN	52821	57821	
158 822	es	**AW**	A	*AW*	MN	52822	57822	
158 823	es	**AW**	A	*AW*	MN	52823	57823	
158 824	es	**AW**	A	*AW*	MN	52824	57824	
158 825	es	**AW**	A	*AW*	MN	52825	57825	
158 826	es	**AW**	A	*AW*	MN	52826	57826	
158 827	es	**AW**	A	*AW*	MN	52827	57827	
158 828	es	**AW**	A	*AW*	MN	52828	57828	
158 829	es	**AW**	A	*AW*	MN	52829	57829	
158 830	es	**AW**	A	*AW*	MN	52830	57830	
158 831	es	**AW**	A	*AW*	MN	52831	57831	
158 832	es	**AW**	A	*AW*	MN	52832	57832	
158 833	es	**AW**	A	*AW*	MN	52833	57833	
158 834	es	**AW**	A	*AW*	MN	52834	57834	
158 835	es	**AW**	A	*AW*	MN	52835	57835	
158 836	es	**AW**	A	*AW*	MN	52836	57836	
158 837	es	**AW**	A	*AW*	MN	52837	57837	
158 838	es	**AW**	A	*AW*	MN	52838	57838	
158 839	es	**AW**	A	*AW*	MN	52839	57839	
158 840	es	**AW**	A	*AW*	MN	52840	57840	

158 841	es	**AW**	A	*AW*	MN	52841	57841
158 842	c	**NO**	A	*NO*	NL	52842	57842
158 843	c	**NO**	A	*NO*	NL	52843	57843
158 844		**NO**	A	*NO*	NL	52844	57844
158 845	v	**NO**	A	*NO*	NL	52845	57845
158 846	††	**ST**	A	*EM*	NM	52846	57846
158 847	††	**ST**	A	*EM*	NM	52847	57847
158 848		**NO**	A	*NO*	NL	52848	57848
158 849	v	**NO**	A	*NO*	NL	52849	57849
158 850	v	**NO**	A	*NO*	NL	52850	57850
158 851		**NO**	A	*NO*	NL	52851	57851
158 852	††	**ST**	A	*EM*	NM	52852	57852
158 853		**NO**	A	*NO*	NL	52853	57853
158 854	†	**ST**	A	*EM*	NM	52854	57854
158 855		**NO**	A	*NO*	NL	52855	57855
158 856	†	**ST**	A	*EM*	NM	52856	57856
158 857	†	**ST**	A	*EM*	NM	52857	57857
158 858	†	**ST**	A	*EM*	NM	52858	57858
158 859		**NO**	A	*NO*	NL	52859	57859
158 860	v	**NO**	A	*NO*	NL	52860	57860
158 861		**NO**	A	*NO*	NL	52861	57861
158 862	†	**ST**	A	*EM*	NM	52862	57862
158 863	†	**ST**	A	*EM*	NM	52863	57863
158 864	†	**ST**	A	*EM*	NM	52864	57864
158 865	†	**ST**	A	*EM*	NM	52865	57865
158 866	†	**ST**	A	*EM*	NM	52866	57866
158 867	c	**SR**	A	*SR*	HA	52867	57867
158 868	c	**SR**	A	*SR*	HA	52868	57868
158 869	c	**SR**	A	*SR*	HA	52869	57869
158 870	c	**SR**	A	*SR*	HA	52870	57870
158 871	c	**SR**	A	*SR*	HA	52871	57871
158 872	c	**NO**	A	*NO*	NL	52872	57872

Names:

158 773	EASTCROFT DEPOT
158 784	Barbara Castle
158 791	County of Nottinghamshire
158 796	Fred Trueman Cricketing Legend
158 797	Jane Tomlinson
158 847	Lincoln Castle Explorer
158 860	Ian Dewhirst
158 861	Magna Carta 800 Lincoln 2015

Class 158/8. Refurbished South West Trains and East Midlands Trains units. Converted from former TransPennine Express units at Wabtec, Doncaster in 2007. 2+1 seating in First Class.

158 885 has been fitted with new ZF transmission as a trial.

* Units refurbished with modifications for operation beyond 2020.

Details as Class 158/0 except:

DMCL. Lot No. 31051 BREL Derby 1989–92. 13/44 1TD 1W (* 13/40(+2) 1TD 1W). 38.5 t.
DMSL. Lot No. 31052 BREL Derby 1989–92. –/70 1T. 38.5 t.

158 880	(158 737)	*	**ST**	P	*SW*	SA	52737	57737
158 881	(158 742)	*	**ST**	P	*SW*	SA	52742	57742
158 882	(158 743)	*	**ST**	P	*SW*	SA	52743	57743
158 883	(158 744)	*	**ST**	P	*SW*	SA	52744	57744
158 884	(158 772)	*	**ST**	P	*SW*	SA	52772	57772
158 885	(158 775)	*	**ST**	P	*SW*	SA	52775	57775
158 886	(158 779)	*	**ST**	P	*SW*	SA	52779	57779
158 887	(158 781)	*	**SW**	P	*SW*	SA	52781	57781
158 888	(158 802)		**ST**	P	*SW*	SA	52802	57802
158 889	(158 808)	*	**ST**	P	*EM*	NM	52808	57808
158 890	(158 814)		**ST**	P	*SW*	SA	52814	57814

CLASS 158/9   BREL

DMSL–DMS. Units leased by West Yorkshire PTE but managed by Eversholt Rail. Details as Class 158/0 except for seating and toilets.

DMSL. Lot No. 31051 BREL Derby 1990–92. –/70 1TD 1W. 38.5 t.
DMS. Lot No. 31052 BREL Derby 1990–92. –/72 and parcels area. 38.5 t.

158 901	**NO**	E	*NO*	NL	52901	57901	
158 902	**NO**	E	*NO*	NL	52902	57902	
158 903	**NO**	E	*NO*	NL	52903	57903	
158 904	**NO**	E	*NO*	NL	52904	57904	
158 905	**NO**	E	*NO*	NL	52905	57905	
158 906	**NO**	E	*NO*	NL	52906	57906	
158 907	**NO**	E	*NO*	NL	52907	57907	
158 908	**NO**	E	*NO*	NL	52908	57908	
158 909	**NO**	E	*NO*	NL	52909	57909	
158 910	**NO**	E	*NO*	NL	52910	57910	William Wilberforce

CLASS 158/0   BREL

DMSL(A)–DMSL(B)–DMSL(A). Units reformed as 3-car hybrid sets for Great Western Railway, mainly used between Cardiff and Portsmouth. For vehicle details see above. Formations can be flexible depending on when unit exams become due.

z – refurbished Great Western Railway units with modifications to comply with the 2020 accessibility regulations.

158 950	z	**GW**	P	*GW*	PM	57751	52761	57761
158 951	z	**GW**	P	*GW*	PM	52751	52764	57764
158 952	w	**FI**	P	*GW*	PM	57745	52762	57762
158 953	w	**FI**	P	*GW*	PM	52745	52750	57750
158 954	w	**FI**	P	*GW*	PM	57747	52760	57760
158 955	w	**FI**	P	*GW*	PM	52747	52765	57765

158 956	z	**GW**	P	*GW*	PM	52748	52768	57768
158 957	z	**GW**	P	*GW*	PM	57748	52771	57771
158 958	w	**FI**	P	*GW*	PM	57746	52776	57776
158 959	w	**FI**	P	*GW*	PM	52746	52778	57778
158 960	w	**FI**	P	*GW*	PM	57749	52769	57769
158 961	z	**GW**	P	*GW*	PM	52749	52767	57767

CLASS 159/0   BREL

DMCL–MSL–DMSL. Built as Class 158. Converted before entering passenger service to Class 159 by Rosyth Dockyard.

Construction: Welded aluminium.
Engines: One Cummins NTA855R3 of 300 kW (400 hp) at 1900 rpm.
Bogies: One BREL P4 (powered) and one BREL T4 (non-powered) per car.
Couplers: BSI.   **Dimensions:** 22.16 x 2.70 m.
Gangways: Throughout.   **Wheel Arrangement:** 2-B + B-2 + B-2.
Doors: Twin-leaf swing plug.   **Maximum Speed:** 90 mph.
Seating Layout: 1: 2+1 facing, 2: 2+2 facing/unidirectional.
Multiple Working: Within class and with Classes 142, 143, 144, 150, 153, 155, 156, 158 and 170.

DMCL. Lot No. 31051 BREL Derby 1992–93. 23/28 1TD 1W. 38.5 t.
MSL. Lot No. 31050 BREL Derby 1992–93. –/70(+6) 1T. 38.5 t.
DMSL. Lot No. 31052 BREL Derby 1992–93. –/72 1T. 38.5 t.

159 001	**ST**	P	*SW*	SA	52873	58718	57873	CITY OF EXETER
159 002	**ST**	P	*SW*	SA	52874	58719	57874	CITY OF SALISBURY
159 003	**ST**	P	*SW*	SA	52875	58720	57875	TEMPLECOMBE
159 004	**ST**	P	*SW*	SA	52876	58721	57876	BASINGSTOKE AND DEANE
159 005	**ST**	P	*SW*	SA	52877	58722	57877	WEST OF ENGLAND LINE
159 006	**ST**	P	*SW*	SA	52878	58723	57878	THE SEATON TRAMWAY Seaton–Colyford–Colyton
159 007	**ST**	P	*SW*	SA	52879	58724	57879	
159 008	**ST**	P	*SW*	SA	52880	58725	57880	
159 009	**ST**	P	*SW*	SA	52881	58726	57881	
159 010	**ST**	P	*SW*	SA	52882	58727	57882	
159 011	**ST**	P	*SW*	SA	52883	58728	57883	
159 012	**ST**	P	*SW*	SA	52884	58729	57884	
159 013	**ST**	P	*SW*	SA	52885	58730	57885	
159 014	**ST**	P	*SW*	SA	52886	58731	57886	
159 015	**ST**	P	*SW*	SA	52887	58732	57887	
159 016	**ST**	P	*SW*	SA	52888	58733	57888	
159 017	**ST**	P	*SW*	SA	52889	58734	57889	
159 018	**ST**	P	*SW*	SA	52890	58735	57890	
159 019	**ST**	P	*SW*	SA	52891	58736	57891	
159 020	**ST**	P	*SW*	SA	52892	58737	57892	
159 021	**ST**	P	*SW*	SA	52893	58738	57893	
159 022	**ST**	P	*SW*	SA	52894	58739	57894	

CLASS 159/1   BREL

DMCL–MSL–DMSL. Units converted from Class 158s at Wabtec, Doncaster in 2006–07 for South West Trains.

Details as Class 158/0 except:
Seating Layout: 1: 2+1 facing, 2: 2+2 facing/unidirectional.

DMCL. Lot No. 31051 BREL Derby 1989–92. 24/24(+2) 1TD 2W. 38.5 t.
MSL. Lot No. 31050 BREL Derby 1989–92. –/70 1T. 38.5 t.
DMSL. Lot No. 31052 BREL Derby 1989–92. –/72 1T. 38.5 t.

159 101	(158 800)	**ST**	P	*SW*	SA	52800	58717	57800
159 102	(158 803)	**ST**	P	*SW*	SA	52803	58703	57803
159 103	(158 804)	**ST**	P	*SW*	SA	52804	58704	57804
159 104	(158 805)	**ST**	P	*SW*	SA	52805	58705	57805
159 105	(158 807)	**ST**	P	*SW*	SA	52807	58707	57807
159 106	(158 809)	**ST**	P	*SW*	SA	52809	58709	57809
159 107	(158 811)	**ST**	P	*SW*	SA	52811	58711	57811
159 108	(158 801)	**ST**	P	*SW*	SA	52801	58701	57801

CLASS 165/0   NETWORK TURBO   BREL

DMSL–DMS and DMSL–MS–DMS. Chiltern Railways units. Refurbished 2003–05 with First Class seats removed and air conditioning fitted.
Construction: Welded aluminium.
Engines: One Perkins 2006-TWH of 260 kW (350 hp) at 2100 rpm.
Bogies: BREL P3-17 (powered), BREL T3-17 (non-powered).
Couplers: BSI.
Dimensions: 23.50/23.25 x 2.81 m.
Gangways: Within unit only.   **Wheel Arrangement:** 2-B (+ B-2) + B-2.
Doors: Twin-leaf swing plug.   **Maximum Speed:** 75 mph.
Seating Layout: 2+2/3+2 facing/unidirectional.
Multiple Working: Within class and with Classes 166, 168, 170 and 172.

Fitted with tripcocks for working over London Underground tracks between Harrow-on-the-Hill and Amersham.

* Refurbished with a new universal access toilet to comply with the 2020 accessibility regulations.

58801–822/58873–878. DMSL. Lot No. 31087 BREL York 1990. –/82(+7) 1T 2W (* –/77(+7) 1TD 2W). 42.1 t.
58823–833. DMSL. Lot No. 31089 BREL York 1991–92. –/82(+7) 1T 2W (* –/77(+7) 1TD 2W). 40.1 t.
MS. Lot No. 31090 BREL York 1991–92. –/106. 37.0 t.
DMS. Lot No. 31088 BREL York 1991–92. –/94. 41.5 t.

165 001		**CR**	A	*CR*	AL	58801	58834
165 002		**CR**	A	*CR*	AL	58802	58835
165 003		**CR**	A	*CR*	AL	58803	58836
165 004	*	**CR**	A	*CR*	AL	58804	58837
165 005		**CR**	A	*CR*	AL	58805	58838
165 006		**CR**	A	*CR*	AL	58806	58839

165 007		**CR**	A	*CR*	AL	58807		58840
165 008	*	**CR**	A	*CR*	AL	58808		58841
165 009		**CR**	A	*CR*	AL	58809		58842
165 010		**CR**	A	*CR*	AL	58810		58843
165 011	*	**CR**	A	*CR*	AL	58811		58844
165 012		**CR**	A	*CR*	AL	58812		58845
165 013		**CR**	A	*CR*	AL	58813		58846
165 014		**CR**	A	*CR*	AL	58814		58847
165 015	*	**CR**	A	*CR*	AL	58815		58848
165 016	*	**CR**	A	*CR*	AL	58816		58849
165 017	*	**CR**	A	*CR*	AL	58817		58850
165 018	*	**CR**	A	*CR*	AL	58818		58851
165 019		**CR**	A	*CR*	AL	58819		58852
165 020	*	**CR**	A	*CR*	AL	58820		58853
165 021	*	**CR**	A	*CR*	AL	58821		58854
165 022		**CR**	A	*CR*	AL	58822		58855
165 023		**CR**	A	*CR*	AL	58873		58867
165 024	*	**CR**	A	*CR*	AL	58874		58868
165 025		**CR**	A	*CR*	AL	58875		58869
165 026	*	**CR**	A	*CR*	AL	58876		58870
165 027		**CR**	A	*CR*	AL	58877		58871
165 028	*	**CR**	A	*CR*	AL	58878		58872
165 029	*	**CR**	A	*CR*	AL	58823	55404	58856
165 030	*	**CR**	A	*CR*	AL	58824	55405	58857
165 031	*	**CR**	A	*CR*	AL	58825	55406	58858
165 032	*	**CR**	A	*CR*	AL	58826	55407	58859
165 033		**CR**	A	*CR*	AL	58827	55408	58860
165 034	*	**CR**	A	*CR*	AL	58828	55409	58861
165 035		**CR**	A	*CR*	AL	58829	55410	58862
165 036	*	**CR**	A	*CR*	AL	58830	55411	58863
165 037		**CR**	A	*CR*	AL	58831	55412	58864
165 038		**CR**	A	*CR*	AL	58832	55413	58865
165 039		**CR**	A	*CR*	AL	58833	55414	58866

CLASS 165/1   NETWORK TURBO   BREL

Great Western Railway units. DMSL–MS–DMS or DMSL–DMS. In 2015 GWR removed First Class from all its Class 165s: all are now Standard Class only.

Construction: Welded aluminium.
Engines: One Perkins 2006-TWH of 260 kW (350 hp) at 2100 rpm.
Bogies: BREL P3-17 (powered), BREL T3-17 (non-powered).
Couplers: BSI.
Dimensions: 23.50/23.25 x 2.81 m.
Gangways: Within unit only.   **Wheel Arrangement:** 2-B (+ B-2) + B-2.
Doors: Twin-leaf swing plug.   **Maximum Speed:** 90 mph.
Seating Layout: 3+2/2+2 facing/unidirectional.
Multiple Working: Within class and with Classes 166, 168, 170 and 172.

* Refurbished with a new universal access toilet to comply with the 2020 accessibility regulations.

58953–969. DMSL. Lot No. 31098 BREL York 1992. –/82 1T (* –/67 1TD 2W).
38.0 t (* 40.8 t).
58879–898. DMSL. Lot No. 31096 BREL York 1992. –/88 1T. 38.0 t.
MS. Lot No. 31099 BREL 1992. –/106. 38.1 t.
DMS. Lot No. 31097 BREL 1992. –/98. 37.0 t.

165 101	**GW**	A	*GW*	RG	58953	55415	58916
165 102	**GW**	A	*GW*	RG	58954	55416	58917
165 103	**GW**	A	*GW*	RG	58955	55417	58918
165 104	*	**GW**	A	*GW*	RG	58956	55418	58919
165 105	*	**GW**	A	*GW*	RG	58957	55419	58920
165 106	*	**GW**	A	*GW*	RG	58958	55420	58921
165 107	*	**GW**	A	*GW*	RG	58959	55421	58922
165 108	*	**GW**	A	*GW*	RG	58960	55422	58923
165 109	*	**GW**	A	*GW*	RG	58961	55423	58924
165 110	*	**GW**	A	*GW*	RG	58962	55424	58925
165 111	*	**GW**	A	*GW*	RG	58963	55425	58926
165 112	*	**GW**	A	*GW*	RG	58964	55426	58927
165 113	*	**GW**	A	*GW*	RG	58965	55427	58928
165 114		**FD**	A	*GW*	RG	58966	55428	58929
165 116		**FD**	A	*GW*	RG	58968	55430	58931
165 117		**FD**	A	*GW*	RG	58969	55431	58932
165 118		**FD**	A	*GW*	RG	58879		58933
165 119		**FD**	A	*GW*	RG	58880		58934
165 120		**FD**	A	*GW*	RG	58881		58935
165 121		**FD**	A	*GW*	RG	58882		58936
165 122		**FD**	A	*GW*	RG	58883		58937
165 123		**FD**	A	*GW*	RG	58884		58938
165 124		**FD**	A	*GW*	RG	58885		58939
165 125		**FD**	A	*GW*	RG	58886		58940
165 126		**FD**	A	*GW*	RG	58887		58941
165 127		**FD**	A	*GW*	RG	58888		58942
165 128		**FD**	A	*GW*	RG	58889		58943
165 129		**FD**	A	*GW*	RG	58890		58944
165 130		**FD**	A	*GW*	RG	58891		58945
165 131		**FD**	A	*GW*	RG	58892		58946
165 132		**FD**	A	*GW*	RG	58893		58947
165 133		**FD**	A	*GW*	RG	58894		58948
165 134		**FD**	A	*GW*	RG	58895		58949
165 135		**FD**	A	*GW*	RG	58896		58950
165 136		**FD**	A	*GW*	RG	58897		58951
165 137		**FD**	A	*GW*	RG	58898		58952

CLASS 166   NETWORK EXPRESS TURBO   ABB

DMCL–MS–DMSL. Great Western Railway units, built for Paddington–
Oxford/Newbury services. Air conditioned and with additional luggage
space compared to the Class 165s. The DMSL vehicles have had their 16
First Class seats declassified. Refurbished with a new universal access
toilet to comply with the 2020 accessibility regulations.

Construction: Welded aluminium.
Engines: One Perkins 2006-TWH of 260 kW (350 hp) at 2100 rpm.
Bogies: BREL P3-17 (powered), BREL T3-17 (non-powered).
Couplers: BSI.
Dimensions: 23.50 x 2.81 m.
Gangways: Within unit only.   **Wheel Arrangement:** 2-B + B-2 + B-2.
Doors: Twin-leaf swing plug.   **Maximum Speed:** 90 mph.
Seating Layout: 1: 2+2 facing, 2: 2+2/3+2 facing/unidirectional.
Multiple Working: Within class and with Classes 165, 168, 170 and 172.

DMCL. Lot No. 31116 ABB York 1992–93. 16/53 1TD 2W. 41.2 t.
MS. Lot No. 31117 ABB York 1992–93. –/91. 39.9 t.
DMSL. Lot No. 31116 ABB York 1992–93. –/84 1T. 39.6 t.

166 201	**FB**	A	*GW*	RG	58101	58601	58122
166 202	**FB**	A	*GW*	RG	58102	58602	58123
166 203	**FB**	A	*GW*	RG	58103	58603	58124
166 204	**GW**	A	*GW*	PM	58104	58604	58125
166 205	**GW**	A	*GW*	PM	58105	58605	58126
166 206	**GW**	A	*GW*	RG	58106	58606	58127
166 207	**FB**	A	*GW*	RG	58107	58607	58128
166 208	**GW**	A	*GW*	PM	58108	58608	58129
166 209	**FB**	A	*GW*	RG	58109	58609	58130
166 210	**GW**	A	*GW*	PM	58110	58610	58131
166 211	**FB**	A	*GW*	RG	58111	58611	58132
166 212	**GW**	A	*GW*	PM	58112	58612	58133
166 213	**GW**	A	*GW*	RG	58113	58613	58134
166 214	**GW**	A	*GW*	PM	58114	58614	58135
166 215	**FB**	A	*GW*	RG	58115	58615	58136
166 216	**GW**	A	*GW*	PM	58116	58616	58137
166 217	**GW**	A	*GW*	RG	58117	58617	58138
166 218	**GW**	A	*GW*	RG	58118	58618	58139
166 219	**GW**	A	*GW*	RG	58119	58619	58140
166 220	**GW**	A	*GW*	RG	58120	58620	58141
166 221	**FB**	A	*GW*	RG	58121	58621	58142

Names:

166 204	Norman Topsom MBE
166 221	Reading Train Care Depot/READING TRAIN CARE DEPOT *(alt sides)*

CLASS 168   CLUBMAN   ADTRANZ/BOMBARDIER

Air conditioned.

Construction: Welded aluminium bodies with bolt-on steel ends.
Engines: One MTU 6R183TD13H of 315 kW (422 hp) at 1900 rpm.
Transmission: Hydraulic. Voith T211rzze to ZF final drive.
Bogies: One Adtranz P3–23 and one BREL T3–23 per car.
Couplers: BSI at outer ends, bar within unit.
Dimensions: Class 168/0: 24.10/23.61 x 2.69 m. Others: 23.62/23.61 x 2.69 m.
Gangways: Within unit only.   **Wheel Arrangement:** 2-B (+ B-2 + B-2) + B-2.
Doors: Twin-leaf swing plug.   **Maximum Speed:** 100 mph.

Seating Layout: 2+2 facing/unidirectional.
Multiple Working: Within class and with Classes 165 and 166.

Fitted with tripcocks for working over London Underground tracks between Harrow-on-the-Hill and Amersham.

Class 168/0. Original Design. DMSL(A)–MS–MSL–DMSL(B) or DMSL(A)–MSL–MS–DMSL(B).

58451–455 were numbered 58656–660 for a time when used in 168 106–110.

58151–155. DMSL(A). Adtranz Derby 1997–98. –/57 1TD 1W. 44.0 t.
58651–655. MSL. Adtranz Derby 1998. –/73 1T. 41.0 t.
58451–455. MS. Adtranz Derby 1998. –/77. 41.0 t.
58251–255. DMSL(B). Adtranz Derby 1998. –/68 1T. 43.6 t.

168 001	**CL**	P	*CR*	AL	58151	58651	58451	58251
168 002	**CL**	P	*CR*	AL	58152	58652	58452	58252
168 003	**CL**	P	*CR*	AL	58153	58453	58653	58253
168 004	**CL**	P	*CR*	AL	58154	58654	58454	58254
168 005	**CL**	P	*CR*	AL	58155	58655	58455	58255

Class 168/1. These units are effectively Class 170s. DMSL(A)–MSL–MS–DMSL(B) or DMSL(A)–MS–DMSL(B).

58461–463 have been renumbered from 58661–663.

58156–163. DMSL(A). Adtranz Derby 2000. –/57 1TD 2W. 45.2 t.
58456–460. MS. Bombardier Derby 2002. –/76. 41.8 t.
58756–757. MSL. Bombardier Derby 2002. –/73 1T. 42.9 t.
58461–463. MS. Adtranz Derby 2000. –/76. 42.4 t.
58256–263. DMSL(B). Adtranz Derby 2000. –/69 1T. 45.2 t.

168 106	**CL**	P	*CR*	AL	58156	58756	58456	58256
168 107	**CL**	P	*CR*	AL	58157	58757	58457	58257
168 108	**CL**	P	*CR*	AL	58158		58458	58258
168 109	**CL**	P	*CR*	AL	58159		58459	58259
168 110	**CL**	P	*CR*	AL	58160		58460	58260
168 111	**CL**	E	*CR*	AL	58161		58461	58261
168 112	**CL**	E	*CR*	AL	58162		58462	58262
168 113	**CL**	E	*CR*	AL	58163		58463	58263

Class 168/2. These units are effectively Class 170s. DMSL(A)–(MS)–MS–DMSL(B).

58164–169. DMSL(A). Bombardier Derby 2003–04. –/57 1TD 2W. 45.4 t.
58365–367. MS. Bombardier Derby 2006. –/76. 43.3 t.
58464/468/469. MS. Bombardier Derby 2003–04. –/76. 44.0 t.
58465–467. MS. Bombardier Derby 2006. –/76. 43.3 t.
58264–269. DMSL(B). Bombardier Derby 2003–04. –/69 1T. 45.5 t.

168 214	**CL**	P	*CR*	AL	58164		58464	58264
168 215	**CL**	P	*CR*	AL	58165	58365	58465	58265
168 216	**CL**	P	*CR*	AL	58166	58466	58366	58266
168 217	**CL**	P	*CR*	AL	58167	58367	58467	58267
168 218	**CL**	P	*CR*	AL	58168		58468	58268
168 219	**CL**	P	*CR*	AL	58169		58469	58269

Class 168/3. Former South West Trains/TransPennine Express Class 170s taken on by Chiltern Railways in 2015–16 and renumbered in the 168 3xx series. 170 309 was originally numbered 170 399. DMSL(A)–DMSL(B).

50301–308/399. DMCL. Adtranz Derby 2000–01. –/59 1TD 2W. 45.8 t.
79301–308/399. DMSL. Adtranz Derby 2000–01. –/69 1T. 45.8 t.

168 321	(170 301)	**CL**	P	*CR*	AL	50301	79301
168 322	(170 302)	**CL**	P	*CR*	AL	50302	79302
168 323	(170 303)	**CL**	P	*CR*	AL	50303	79303
168 324	(170 304)	**CL**	P	*CR*	AL	50304	79304
168 325	(170 305)	**CL**	P	*CR*	AL	50305	79305
168 326	(170 306)	**CL**	P	*CR*	AL	50306	79306
168 327	(170 307)	**CL**	P	*CR*	AL	50307	79307
168 328	(170 308)	**CL**	P	*CR*	AL	50308	79308
168 329	(170 309)	**CL**	P	*CR*	AL	50399	79399

CLASS 170 TURBOSTAR ADTRANZ/BOMBARDIER

Various formations. Air conditioned.

Construction: Welded aluminium bodies with bolt-on steel ends.
Engines: One MTU 6R183TD13H of 315 kW (422 hp) at 1900 rpm.
Transmission: Hydraulic. Voith T211rzze to ZF final drive.
Bogies: One Adtranz P3–23 and one BREL T3–23 per car.
Couplers: BSI at outer ends, bar within later build units.
Dimensions: 23.62/23.61 x 2.69 m.
Gangways: Within unit only.   **Wheel Arrangement:** 2-B (+ B-2) + B-2.
Doors: Twin-leaf sliding plug.   **Maximum Speed:** 100 mph.
Seating Layout: 1: 2+1 facing/unidirectional. 2: 2+2 unidirectional/facing.
Multiple Working: Within class and with Classes 150, 153, 155, 156, 158, 159 and 172.

Class 170/1. CrossCountry (former Midland Mainline) units. Lazareni seating. DMSL–MS–DMCL/DMSL–DMCL.

DMSL. Adtranz Derby 1998–99. –/59 1TD 2W. 45.0 t.
MS. Adtranz Derby 2001. –/80. 43.0 t.
DMCL. Adtranz Derby 1998–99. 9/52 1T. 44.8 t

170 101	**XC**	P	*XC*	TS	50101	55101	79101
170 102	**XC**	P	*XC*	TS	50102	55102	79102
170 103	**XC**	P	*XC*	TS	50103	55103	79103
170 104	**XC**	P	*XC*	TS	50104	55104	79104
170 105	**XC**	P	*XC*	TS	50105	55105	79105
170 106	**XC**	P	*XC*	TS	50106	55106	79106
170 107	**XC**	P	*XC*	TS	50107	55107	79107
170 108	**XC**	P	*XC*	TS	50108	55108	79108
170 109	**XC**	P	*XC*	TS	50109	55109	79109
170 110	**XC**	P	*XC*	TS	50110	55110	79110
170 111	**XC**	P	*XC*	TS	50111		79111
170 112	**XC**	P	*XC*	TS	50112		79112
170 113	**XC**	P	*XC*	TS	50113		79113
170 114	**XC**	P	*XC*	TS	50114		79114

170 115	**XC**	P	*XC*	TS	50115	79115
170 116	**XC**	P	*XC*	TS	50116	79116
170 117	**XC**	P	*XC*	TS	50117	79117

Class 170/2. Greater Anglia 3-car units. Chapman seating. DMCL–MSL–DMSL.

DMCL. Adtranz Derby 1999. 7/39 1TD 2W. 45.0 t.
MSL. Adtranz Derby 1999. –/68 1T. Guard's office. 45.3 t.
DMSL. Adtranz Derby 1999. –/66 1T. 43.4 t.

170 201	r	**GA**	P	*GA*	NC	50201	56201	79201
170 202	r	**1**	P	*GA*	NC	50202	56202	79202
170 203	r	**GA**	P	*GA*	NC	50203	56203	79203
170 204	r	**1**	P	*GA*	NC	50204	56204	79204
170 205	r	**1**	P	*GA*	NC	50205	56205	79205
170 206	r	**GA**	P	*GA*	NC	50206	56206	79206
170 207	r	**1**	P	*GA*	NC	50207	56207	79207
170 208	r	**GA**	P	*GA*	NC	50208	56208	79208

Class 170/2. Greater Anglia 2-car units. Chapman seating. DMSL–DMCL.

DMSL. Bombardier Derby 2002. –/57 1TD 2W. 45.7 t.
DMCL. Bombardier Derby 2002. 9/53 1T. 45.7 t.

170 270	r	**GA**	P	*GA*	NC	50270	79270
170 271	r	**AN**	P	*GA*	NC	50271	79271
170 272	r	**AN**	P	*GA*	NC	50272	79272
170 273	r	**AN**	P	*GA*	NC	50273	79273

Class 170/3. Units built for Hull Trains, now in use with ScotRail. Chapman seating. DMSL–MSLRB–DMSL.

DMSL(A). Bombardier Derby 2004. –/55 1TD 2W. 46.5 t.
MSLRB. Bombardier Derby 2004. –/57 1T. Buffet and guard's office 44.7 t.
DMSL(B). Bombardier Derby 2004. –/67 1T. 47.0 t.

170 393	**SR**	P	*SR*	HA	50393	56393	79393
170 394	**SR**	P	*SR*	HA	50394	56394	79394
170 395	**SR**	P	*SR*	HA	50395	56395	79395
170 396	**SR**	P	*SR*	HA	50396	56396	79396

Class 170/3. CrossCountry units. Lazareni seating. DMSL–MS–DMCL.

DMSL. Bombardier Derby 2002. –/59 1TD 2W. 45.4 t.
MS. Bombardier Derby 2002. –/80. 43.0 t.
DMCL. Bombardier Derby 2002. 9/52 1T. 45.8 t.

| 170 397 | **XC** | P | *XC* | TS | 50397 | 56397 | 79397 |
| 170 398 | **XC** | P | *XC* | TS | 50398 | 56398 | 79398 |

Class 170/4. ScotRail "express" units. Chapman seating. DMCL–MS–DMCL.

170 416–420 are sub-leased from Southern to ScotRail.

170 421–424 have been renumbered as Southern 171 201/202 and 171 401/402.

Non-standard/Advertising liveries:

170 407   BTP text number 61016 (blue).
170 416, 170 417, 170 419, 170 420 Unbranded Saltire blue with grey doors.

DMCL(A). Adtranz Derby 1999–2001. 9/43 1TD 2W. 45.2 t.
MS. Adtranz Derby 1999–2001. –/76. 42.5 t.
DMCL(B). Adtranz Derby 1999–2001. 9/49 1T. 45.2 t.

170 401	**SR**	P	SR	HA	50401	56401	79401
170 402	**SR**	P	SR	HA	50402	56402	79402
170 403	**SR**	P	SR	HA	50403	56403	79403
170 404	**SR**	P	SR	HA	50404	56404	79404
170 405	**SR**	P	SR	HA	50405	56405	79405
170 406	**SR**	P	SR	HA	50406	56406	79406
170 407	**AL**	P	SR	HA	50407	56407	79407
170 408	**SR**	P	SR	HA	50408	56408	79408
170 409	**SR**	P	SR	HA	50409	56409	79409
170 410	**SR**	P	SR	HA	50410	56410	79410
170 411	**SR**	P	SR	HA	50411	56411	79411
170 412	**SR**	P	SR	HA	50412	56412	79412
170 413	**SR**	P	SR	HA	50413	56413	79413
170 414	**SR**	P	SR	HA	50414	56414	79414
170 415	**SR**	P	SR	HA	50415	56415	79415
170 416	**0**	E	SR	HA	50416	56416	79416
170 417	**0**	E	SR	HA	50417	56417	79417
170 418	**SR**	E	SR	HA	50418	56418	79418
170 419	**0**	E	SR	HA	50419	56419	79419
170 420	**0**	E	SR	HA	50420	56420	79420

Class 170/4. ScotRail "express" units. Chapman seating. DMCL–MS–DMCL.

DMCL. Bombardier Derby 2003–05. 9/43 1TD 2W. 46.8 t.
MS. Bombardier Derby 2003–05. –/76. 43.7 t.
DMCL. Bombardier Derby 2003–05. 9/49 1T. 46.5 t.

170 425	**SR**	P	SR	HA	50425	56425	79425
170 426	**SR**	P	SR	HA	50426	56426	79426
170 427	**SR**	P	SR	HA	50427	56427	79427
170 428	**SR**	P	SR	HA	50428	56428	79428
170 429	**SR**	P	SR	HA	50429	56429	79429
170 430	**SR**	P	SR	HA	50430	56430	79430
170 431	**SR**	P	SR	HA	50431	56431	79431
170 432	**SR**	P	SR	HA	50432	56432	79432
170 433	**SR**	P	SR	HA	50433	56433	79433
170 434	**SR**	P	SR	HA	50434	56434	79434

Class 170/4. ScotRail units. Originally built as Standard Class only units. 170 450–457 have been retro-fitted with First Class. Chapman seating. DMSL–MS–DMSL or † DMCL–MS–DMCL.

DMSL. Bombardier Derby 2004–05. –/55 1TD 2W († 9/47 1TD 2W). 46.3 t.
MS. Bombardier Derby 2004–05. –/76. 43.4 t.
DMSL. Bombardier Derby 2004–05. –/67 1T († 9/49 1T 1W). 46.4 t.

170 450	†	**SR**	P	SR	HA	50450	56450	79450
170 451	†	**SR**	P	SR	HA	50451	56451	79451
170 452	†	**SR**	P	SR	HA	50452	56452	79452
170 453	†	**SR**	P	SR	HA	50453	56453	79453

170 454	†	**SR**	P	*SR*	HA	50454	56454	79454
170 455	†	**SR**	P	*SR*	HA	50455	56455	79455
170 456	†	**SR**	P	*SR*	HA	50456	56456	79456
170 457	†	**SR**	P	*SR*	HA	50457	56457	79457
170 458		**SR**	P	*SR*	HA	50458	56458	79458
170 459		**SR**	P	*SR*	HA	50459	56459	79459
170 460		**SR**	P	*SR*	HA	50460	56460	79460
170 461		**SR**	P	*SR*	HA	50461	56461	79461

Class 170/4. ScotRail units. Standard Class only units. Chapman seating. DMSL–MS–DMSL.

50470–471. DMSL(A). Adtranz Derby 2001. –/55 1TD 2W. 45.1 t.
50472–478. DMSL(A). Bombardier Derby 2004–05. –/57 1TD 2W. 46.3 t.
56470–471. MS. Adtranz Derby 2001. –/76. 42.4 t.
56472–478. MS. Bombardier Derby 2004–05. –/76. 43.4 t.
79470–471. DMSL(B). Adtranz Derby 2001. –/67 1T. 45.1 t.
79472–478. DMSL(B). Bombardier Derby 2004–05. –/67 1T. 46.4 t.

170 470	**SR**	P	*SR*	HA	50470	56470	79470
170 471	**SR**	P	*SR*	HA	50471	56471	79471
170 472	**SR**	P	*SR*	HA	50472	56472	79472
170 473	**SR**	P	*SR*	HA	50473	56473	79473
170 474	**SR**	P	*SR*	HA	50474	56474	79474
170 475	**SR**	P	*SR*	HA	50475	56475	79475
170 476	**SR**	P	*SR*	HA	50476	56476	79476
170 477	**SR**	P	*SR*	HA	50477	56477	79477
170 478	**SR**	P	*SR*	HA	50478	56478	79478

Class 170/5. London Midland and CrossCountry 2-car units. Lazareni seating. DMSL–DMSL or * DMSL–DMCL (CrossCountry).

DMSL(A). Adtranz Derby 1999–2000. –/55 1TD 2W (* –/59 1TD 2W). 45.8 t.
DMSL(B). Adtranz Derby 1999–2000. –/67 1T (* DMCL 9/52 1T). 45.9 t.

170 501		**LM**	P	*LM*	TS	50501	79501
170 502		**LM**	P	*LM*	TS	50502	79502
170 503		**LM**	P	*LM*	TS	50503	79503
170 504		**LM**	P	*LM*	TS	50504	79504
170 505		**LM**	P	*LM*	TS	50505	79505
170 506		**LM**	P	*LM*	TS	50506	79506
170 507		**LM**	P	*LM*	TS	50507	79507
170 508		**LM**	P	*LM*	TS	50508	79508
170 509		**LM**	P	*LM*	TS	50509	79509
170 510		**LM**	P	*LM*	TS	50510	79510
170 511		**LM**	P	*LM*	TS	50511	79511
170 512		**LM**	P	*LM*	TS	50512	79512
170 513		**LM**	P	*LM*	TS	50513	79513
170 514		**LM**	P	*LM*	TS	50514	79514
170 515		**LM**	P	*LM*	TS	50515	79515
170 516		**LM**	P	*LM*	TS	50516	79516
170 517		**LM**	P	*LM*	TS	50517	79517
170 518	*	**XC**	P	*XC*	TS	50518	79518
170 519	*	**XC**	P	*XC*	TS	50519	79519

170 520	*	**XC**	P	*XC*	TS	50520	79520
170 521	*	**XC**	P	*XC*	TS	50521	79521
170 522	*	**XC**	P	*XC*	TS	50522	79522
170 523	*	**XC**	P	*XC*	TS	50523	79523

Class 170/6. London Midland and CrossCountry 3-car units. Lazareni seating. DMSL–MS–DMSL or * DMSL–MS–DMCL (CrossCountry).

DMSL(A). Adtranz Derby 2000. –/55 1TD 2W (* –/59 1TD 2W). 45.8 t.
MS. Adtranz Derby 2000. –/74 (* –/80). 42.4 t.
DMSL(B). Adtranz Derby 2000. –/67 1T (* DMCL 9/52 1T). 45.9 t.

170 630		**LM**	P	*LM*	TS	50630	56630	79630
170 631		**LM**	P	*LM*	TS	50631	56631	79631
170 632		**LM**	P	*LM*	TS	50632	56632	79632
170 633		**LM**	P	*LM*	TS	50633	56633	79633
170 634		**LM**	P	*LM*	TS	50634	56634	79634
170 635		**LM**	P	*LM*	TS	50635	56635	79635
170 636	*	**XC**	P	*XC*	TS	50636	56636	79636
170 637	*	**XC**	P	*XC*	TS	50637	56637	79637
170 638	*	**XC**	P	*XC*	TS	50638	56638	79638
170 639	*	**XC**	P	*XC*	TS	50639	56639	79639

CLASS 171   TURBOSTAR   BOMBARDIER

DMCL–DMSL or DMCL–MS–MS–DMCL. Southern units. Air conditioned. Chapman seating.

Construction: Welded aluminium bodies with bolt-on steel ends.
Engines: One MTU 6R183TD13H of 315 kW (422 hp) at 1900 rpm.
Transmission: Hydraulic. Voith T211rzze to ZF final drive.
Bogies: One Adtranz P3–23 and one BREL T3–23 per car.
Couplers: Dellner 12 at outer ends, bar within unit (Class 171/8).
Dimensions: 23.62/23.61 x 2.69 m.
Gangways: Within unit only.   **Wheel Arrangement:** 2-B (+ B-2 + B-2) + B-2.
Doors: Twin-leaf swing plug.   **Maximum Speed:** 100 mph.
Seating Layout: 1: 2+1 facing/unidirectional. 2: 2+2 facing/unidirectional.
Multiple Working: Within class and with EMU Classes 375 and 377 in an emergency.

Class 171/2. 2-car units rebuilt from ScotRail Class 170s. Full details awaited. DMCL–DMSL.

Originally built as 3-car units 170 421/423, but renumbered as Class 171 on fitting with Dellner couplers.

DMCL. Adtranz Derby 1999–2001.
DMSL. Adtranz Derby 1999–2001.

| 171 201 | **SN** | E | *SN* | SU | 50421 | 79421 |
| 171 202 | **SN** | E | *SN* | SU | 50423 | 79423 |

Class 171/4. 4-car units rebuilt from ScotRail Class 170s. Full details awaited. DMCL(A)–MS–MS–DMCL(B).

Reformed and renumbered Class 171s in 2016 using vehicles from ScotRail 3-car Class 170s 170 421–424.

DMCL(A). Adtranz Derby 1999–2001.
MS. Adtranz Derby 1999–2001.
DMCL(B). Adtranz Derby 1999–2001.

| 171 401 | **SN** | E | *SN* | SU | 50422 | 56421 | 56422 | 79422 |
| 171 402 | **SN** | E | *SN* | SU | 50424 | 56423 | 56424 | 79424 |

Class 171/7. 2-car units. DMCL–DMSL.

171 721–726 were built as Class 170s (170 721–726), but renumbered as Class 171 on fitting with Dellner couplers.

171 730 was formerly South West Trains unit 170 392, before transferring to Southern in 2007.

50721–726. DMCL. Bombardier Derby 2003. 9/43 1TD 2W. 47.6 t.
50727–729. DMCL. Bombardier Derby 2005. 9/43 1TD 2W. 46.3 t.
50392. DMCL. Bombardier Derby 2003. 9/43 1TD 2W. 46.6 t.
79721–726. DMSL. Bombardier Derby 2003. –/64 1T. 47.8 t.
79727–729. DMSL. Bombardier Derby 2005. –/64 1T. 46.2 t.
79392. DMSL. Bombardier Derby 2003. –/64 1T. 46.5 t.

171 721	**SN**	P	*SN*	SU	50721	79721
171 722	**SN**	P	*SN*	SU	50722	79722
171 723	**SN**	P	*SN*	SU	50723	79723
171 724	**SN**	P	*SN*	SU	50724	79724
171 725	**SN**	P	*SN*	SU	50725	79725
171 726	**SN**	P	*SN*	SU	50726	79726
171 727	**SN**	P	*SN*	SU	50727	79727
171 728	**SN**	P	*SN*	SU	50728	79728
171 729	**SN**	P	*SN*	SU	50729	79729
171 730	**SN**	P	*SN*	SU	50392	79392

Class 171/8. 4-car units. DMCL(A)–MS–MS–DMCL(B).

DMCL(A). Bombardier Derby 2004. 9/43 1TD 2W. 46.5 t.
MS. Bombardier Derby 2004. –/74. 43.7 t.
DMCL(B). Bombardier Derby 2004. 9/50 1T. 46.5 t.

171 801	**SN**	P	*SN*	SU	50801	54801	56801	79801
171 802	**SN**	P	*SN*	SU	50802	54802	56802	79802
171 803	**SN**	P	*SN*	SU	50803	54803	56803	79803
171 804	**SN**	P	*SN*	SU	50804	54804	56804	79804
171 805	**SN**	P	*SN*	SU	50805	54805	56805	79805
171 806	**SN**	P	*SN*	SU	50806	54806	56806	79806

CLASS 172

New generation Lon~~don~~ **TURBOSTAR** **BOM~~BARDIER~~**
Midland Turbostars. Air ~~conditioned,~~ Chiltern Railways and Lon~~don~~

Construction: Welded alumin~~ium~~ ~~bod~~y, ~~with bolt-on steel ends.~~
Engines: One MTU 6H1800R83 o~~f~~ ~~(3 hp) at 1800 rpm.~~
Transmission: Mechanical. Supplie~~d~~ ~~with bolt-on steel ends.~~
Bogies: B5006 type "lightweight" bogie~~s~~ ~~(3 hp) at 1800 rpm.~~
Couplers: BSI at outer ends, bar within uni~~t.~~ ~~many.~~
Dimensions: 23.62/23.0 x 2.69 m.
Gangways: London Overground & Chiltern units. ~~With~~ ~~in unit only. London~~
Midland units: Throughout.
Wheel Arrangement: 2-B (+ B-2) + B-2.
Doors: Twin-leaf sliding plug.
Maximum Speed: 100 mph (London Overground units 75 mph).
Seating Layout: 2+2 facing/unidirectional.
Multiple Working: Within class and with Classes 150, 153, 155, 156, 158, 159, 165, 166 and 170.

Class 172/0. London Overground units. Used on the Gospel Oak–Barking line. DMS–DMS.

59311–318. DMS(W). Bombardier Derby 2009–10. –/60 2W. 41.6 t.
59411–418. DMS. Bombardier Derby 2009–10. –/64. 41.5 t.

172 001	LO	A	LO	WN	59311 59411
172 002	LO	A	LO	WN	59312 59412
172 003	LO	A	LO	WN	59313 59413
172 004	LO	A	LO	WN	59314 59414
172 005	LO	A	LO	WN	59315 59415
172 006	LO	A	LO	WN	59316 59416
172 007	LO	A	LO	WN	59317 59417
172 008	LO	A	LO	WN	59318 59418

Class 172/1. Chiltern Railways units. DMSL–DMS.

59111–114. DMSL. Bombardier Derby 2009–10. –/60(+5) 1TD 2W. 42.4 t.
59211–214. DMS. Bombardier Derby 2009–10. –/80. 41.8 t.

172 101	CR	A	CR	AL	59111 59211
172 102	CR	A	CR	AL	59112 59212
172 103	CR	A	CR	AL	59113 59213
172 104	CR	A	CR	AL	59114 59214

Class 172/2. London Midland 2-car units. DMSL–DMS. Used on local services via Birmingham Snow Hill.

50211–222. DMSL. Bombardier Derby 2010–11. –/52(+11) 1TD 2W. 42.5 t.
79211–222. DMS. Bombardier Derby 2010–11. –/68(+8). 41.9 t.

172 211	LM	P	LM	TS	50211 79211
172 212	LM	P	LM	TS	50212 79212
172 213	LM	P	LM	TS	50213 79213
172 214	LM	P	LM	TS	50214 79214

							79215
	LM	P	LM		TS		79216
	LM	P	LM		TS		79217
	LM	P	LM		TS	219	79218
	LM	P	LM		50220		79219
	LM	P	LM		50221		79220
	LM	P	LM		50222		79221
	LM	P	LM				79222
	LM	P	LM				

3-car units. DMSL–MS–DMS. Used on local services via Birmin Snow Hill.

Class 172/3. London

50331–345. DMS. Bombardier Derby 2010–11. –/52(+11) 1TD 2W. 42.5 t.
56331–345. MS. Bombardier Derby 2010–11. –/72(+8). 38.8 t.
79331–345. D Bombardier Derby 2010–11. –/68(+8). 41.9 t.

172 331	LM	P	LM	TS	50331	56331	79331
172 332	LM	P	LM	TS	50332	56332	79332
172 333	LM	P	LM	TS	50333	56333	79333
172 334	LM	P	LM	TS	50334	56334	79334
172 335	LM	P	LM	TS	50335	56335	79335
172 336	LM	P	LM	TS	50336	56336	79336
172 337	LM	P	LM	TS	50337	56337	79337
172 338	LM	P	LM	TS	50338	56338	79338
172 339	LM	P	LM	TS	50339	56339	79339
172 340	LM	P	LM	TS	50340	56340	79340
172 341	LM	P	LM	TS	50341	56341	79341
172 342	LM	P	LM	TS	50342	56342	79342
172 343	LM	P	LM	TS	50343	56343	79343
172 344	LM	P	LM	TS	50344	56344	79344
172 345	LM	P	LM	TS	50345	56345	79345

CLASS 175 CORADIA 1000 ALSTOM

Air conditioned.

Construction: Steel.
Engines: One Cummins N14 of 335 kW (450 hp).
Transmission: Hydraulic. Voith T211rzze to ZF Voith final drive.
Bogies: ACR (Alstom FBO) – LTB-MBS1, TB-MB1, MBS1-LTB.
Couplers: Scharfenberg outer ends and bar within unit (Class 175/1).
Dimensions: 23.7 x 2.73 m.
Gangways: Within unit only. **Wheel Arrangement:** 2-B (+ B-2) + B-2.
Doors: Single-leaf swing plug. **Maximum Speed:** 100 mph.
Seating Layout: 2+2 facing/unidirectional.
Multiple Working: Within class and with Class 180.

175 004 and 175 005 are operating in misformed formations at the time of writing.

Class 175/0. DMSL–DMSL. 2-car units.

DMSL(A). Alstom Birmingham 1999–2000. –/54 1TD 2W. 48.8 t.
DMSL(B). Alstom Birmingham 1999–2000. –/64 1T. 50.7 t.

175 001	**AV**	A	*AW*	CH	50701	79701
175 002	**AV**	A	*AW*	CH	50702	79702
175 003	**AV**	A	*AW*	CH	50703	79703
175 004	**AV**	A	*AW*	CH	50705	79704
175 005	**AV**	A	*AW*	CH	50704	79705
175 006	**AV**	A	*AW*	CH	50706	79706
175 007	**AV**	A	*AW*	CH	50707	79707
175 008	**AV**	A	*AW*	CH	50708	79708
175 009	**AV**	A	*AW*	CH	50709	79709
175 010	**AV**	A	*AW*	CH	50710	79710
175 011	**AV**	A	*AW*	CH	50711	79711

Class 175/1. DMSL–MSL–DMSL. 3-car units.

DMSL(A). Alstom Birmingham 1999–2001. –/54 1TD 2W. 50.7 t.
MSL. Alstom Birmingham 1999–2001. –/68 1T. 47.5 t.
DMSL(B). Alstom Birmingham 1999–2001. –/64 1T. 49.5 t.

175 101	**AV**	A	*AW*	CH	50751	56751	79751
175 102	**AV**	A	*AW*	CH	50752	56752	79752
175 103	**AV**	A	*AW*	CH	50753	56753	79753
175 104	**AV**	A	*AW*	CH	50754	56754	79754
175 105	**AV**	A	*AW*	CH	50755	56755	79755
175 106	**AV**	A	*AW*	CH	50756	56756	79756
175 107	**AV**	A	*AW*	CH	50757	56757	79757
175 108	**AV**	A	*AW*	CH	50758	56758	79758
175 109	**AV**	A	*AW*	CH	50759	56759	79759
175 110	**AV**	A	*AW*	CH	50760	56760	79760
175 111	**AV**	A	*AW*	CH	50761	56761	79761
175 112	**AV**	A	*AW*	CH	50762	56762	79762
175 113	**AV**	A	*AW*	CH	50763	56763	79763
175 114	**AV**	A	*AW*	CH	50764	56764	79764
175 115	**AV**	A	*AW*	CH	50765	56765	79765
175 116	**AV**	A	*AW*	CH	50766	56766	79766

CLASS 180   CORADIA 1000   ALSTOM

Air conditioned.

Construction: Steel.
Engines: One Cummins QSK19 of 560 kW (750 hp) at 2100 rpm.
Transmission: Hydraulic. Voith T312br to Voith final drive.
Bogies: ACR (Alstom FBO): LTB1-MBS2, TB1-MB2, TB1-MB2, TB2-MB2, MBS2-LTB1.
Couplers: Scharfenberg outer ends, bar within unit.
Dimensions: 23.71/23.03 x 2.73 m.
Gangways: Within unit only.
Wheel Arrangement: 2-B + B-2 + B-2 + B-2 + B-2.
Doors: Single-leaf swing plug.   **Maximum Speed:** 125 mph.
Seating Layout: 1: 2+1 facing/unidirectional, 2: 2+2 facing/unidirectional.
Multiple Working: Within class and with Class 175.

DMSL(A). Alstom Birmingham 2000–01. –/46 2W 1TD. 51.7 t.
MFL. Alstom Birmingham 2000–01. 42/– 1T 1W + catering point. 49.6 t.
MSL. Alstom Birmingham 2000–01. –/68 1T. 49.5 t.
MSLRB. Alstom Birmingham 2000–01. –/56 1T. 50.3 t.
DMSL(B). Alstom Birmingham 2000–01. –/56 1T. 51.4 t.

180 101	**GC**	A	*GC*	HT	50901	54901	55901	56901	59901
180 102	**FD**	A	*GW*	OO	50902	54902	55902	56902	59902
180 103	**FD**	A	*GW*	OO	50903	54903	55903	56903	59903
180 104	**FD**	A	*GW*	OO	50904	54904	55904	56904	59904
180 105	**GC**	A	*GC*	HT	50905	54905	55905	56905	59905
180 106	**FD**	A	*GW*	OO	50906	54906	55906	56906	59906
180 107	**GC**	A	*GC*	HT	50907	54907	55907	56907	59907
180 108	**FD**	A	*GC*	HT	50908	54908	55908	56908	59908
180 109	**FD**	A	*HT*	OO	50909	54909	55909	56909	59909
180 110	**FD**	A	*HT*	OO	50910	54910	55910	56910	59910
180 111	**FD**	A	*HT*	OO	50911	54911	55911	56911	59911
180 112	**GC**	A	*GC*	HT	50912	54912	55912	56912	59912
180 113	**FD**	A	*HT*	OO	50913	54913	55913	56913	59913
180 114	**GC**	A	*GC*	HT	50914	54914	55914	56914	59914

Names (carried on DMSL(A):

180 105	THE YORKSHIRE ARTIST ASHLEY JACKSON
180 107	HART OF THE NORTH
180 112	JAMES HERRIOT
180 114	KIRKGATE CALLING

CLASS 185   DESIRO UK   SIEMENS

Air conditioned. Grammer seating.

Construction: Aluminium.
Engines: One Cummins QSK19 of 560 kW (750 hp) at 2100 rpm.
Transmission: Voith.
Bogies: Siemens.
Couplers: Dellner 12.
Gangways: Within unit only.
Doors: Double-leaf sliding plug.
Seating Layout: 1: 2+1 facing/unidirectional, 2: 2+2 facing/unidirectional.
Multiple Working: Within class only.

Dimensions: 23.76/23.75 x 2.66 m.
Wheel Arrangement: 2-B + 2-B + B-2.
Maximum Speed: 100 mph.

DMCL. Siemens Krefeld 2005–06. 15/18(+8) 2W 1TD + catering point. 55.4 t.
MSL. Siemens Krefeld 2005–06. –/72 1T. 52.7 t.
DMS. Siemens Krefeld 2005–06. –/64(4). 54.9 t.

185 101	**TP**	E	*TP*	AK	51101	53101	54101
185 102	**TP**	E	*TP*	AK	51102	53102	54102
185 103	**TP**	E	*TP*	AK	51103	53103	54103
185 104	**TP**	E	*TP*	AK	51104	53104	54104
185 105	**TP**	E	*TP*	AK	51105	53105	54105
185 106	**TP**	E	*TP*	AK	51106	53106	54106
185 107	**TP**	E	*TP*	AK	51107	53107	54107

185 108	**TP**	E	*TP*	AK	51108	53108	54108
185 109	**TP**	E	*TP*	AK	51109	53109	54109
185 110	**TP**	E	*TP*	AK	51110	53110	54110
185 111	**TP**	E	*TP*	AK	51111	53111	54111
185 112	**TP**	E	*TP*	AK	51112	53112	54112
185 113	**TP**	E	*TP*	AK	51113	53113	54113
185 114	**TP**	E	*TP*	AK	51114	53114	54114
185 115	**TP**	E	*TP*	AK	51115	53115	54115
185 116	**TP**	E	*TP*	AK	51116	53116	54116
185 117	**TP**	E	*TP*	AK	51117	53117	54117
185 118	**TP**	E	*TP*	AK	51118	53118	54118
185 119	**TP**	E	*TP*	AK	51119	53119	54119
185 120	**TP**	E	*TP*	AK	51120	53120	54120
185 121	**TP**	E	*TP*	AK	51121	53121	54121
185 122	**TP**	E	*TP*	AK	51122	53122	54122
185 123	**TP**	E	*TP*	AK	51123	53123	54123
185 124	**TP**	E	*TP*	AK	51124	53124	54124
185 125	**TP**	E	*TP*	AK	51125	53125	54125
185 126	**TP**	E	*TP*	AK	51126	53126	54126
185 127	**TP**	E	*TP*	AK	51127	53127	54127
185 128	**TP**	E	*TP*	AK	51128	53128	54128
185 129	**TP**	E	*TP*	AK	51129	53129	54129
185 130	**TP**	E	*TP*	AK	51130	53130	54130
185 131	**TP**	E	*TP*	AK	51131	53131	54131
185 132	**TP**	E	*TP*	AK	51132	53132	54132
185 133	**TP**	E	*TP*	AK	51133	53133	54133
185 134	**TP**	E	*TP*	AK	51134	53134	54134
185 135	**TP**	E	*TP*	AK	51135	53135	54135
185 136	**TP**	E	*TP*	AK	51136	53136	54136
185 137	**TP**	E	*TP*	AK	51137	53137	54137
185 138	**TP**	E	*TP*	AK	51138	53138	54138
185 139	**TP**	E	*TP*	AK	51139	53139	54139
185 140	**TP**	E	*TP*	AK	51140	53140	54140
185 141	**TP**	E	*TP*	AK	51141	53141	54141
185 142	**TP**	E	*TP*	AK	51142	53142	54142
185 143	**TP**	E	*TP*	AK	51143	53143	54143
185 144	**TP**	E	*TP*	AK	51144	53144	54144
185 145	**TP**	E	*TP*	AK	51145	53145	54145
185 146	**TP**	E	*TP*	AK	51146	53146	54146
185 147	**TP**	E	*TP*	AK	51147	53147	54147
185 148	**TP**	E	*TP*	AK	51148	53148	54148
185 149	**TP**	E	*TP*	AK	51149	53149	54149
185 150	**TP**	E	*TP*	AK	51150	53150	54150
185 151	**TP**	E	*TP*	AK	51151	53151	54151

CLASS 195   CIVITY   CAF

DMS–DMS or DMS–MS–DMS. New units under construction for Northern, for delivery 2018–19. Air conditioned. Full details awaited.

Construction: Aluminium.
Engines: One Rolls-Royce MTU 6H 1800 R85L of 390 kW (523 hp) per car.
Transmission: ZF.
Bogies: CAF.

Couplers: Dellner.	**Dimensions:**
Gangways: Within unit only.	**Wheel Arrangement:**
Doors:	**Maximum Speed:** 100 mph.

Seating Layout:
Multiple Working: Within class only.

Class 195/0. DMS–DMS. 2-car units.

DMS(A). CAF Irun 2017–19.
DMS(B). CAF Irun 2017–19.

195 001	E	101001	103001
195 002	E	101002	103002
195 003	E	101003	103003
195 004	E	101004	103004
195 005	E	101005	103005
195 006	E	101006	103006
195 007	E	101007	103007
195 008	E	101008	103008
195 009	E	101009	103009
195 010	E	101010	103010
195 011	E	101011	103011
195 012	E	101012	103012
195 013	E	101013	103013
195 014	E	101014	103014
195 015	E	101015	103015
195 016	E	101016	103016
195 017	E	101017	103017
195 018	E	101018	103018
195 019	E	101019	103019
195 020	E	101020	103020
195 021	E	101021	103021
195 022	E	101022	103022
195 023	E	101023	103023
195 024	E	101024	103024
195 025	E	101025	103025

Class 195/1. DMS–MS–DMS. 3-car units.

DMS(A). CAF Irun 2017–19.
MS. CAF Irun 2017–19.
DMS(B). CAF Irun 2017–19.

195 101	E	101101	102101	103101
195 102	E	101102	102102	103102

195 103	E	101103	102103	103103
195 104	E	101104	102104	103104
195 105	E	101105	102105	103105
195 106	E	101106	102106	103106
195 107	E	101107	102107	103107
195 108	E	101108	102108	103108
195 109	E	101109	102109	103109
195 110	E	101110	102110	103110
195 111	E	101111	102111	103111
195 112	E	101112	102112	103112
195 113	E	101113	102113	103113
195 114	E	101114	102114	103114
195 115	E	101115	102115	103115
195 116	E	101116	102116	103116
195 117	E	101117	102117	103117
195 118	E	101118	102118	103118
195 119	E	101119	102119	103119
195 120	E	101120	102120	103120
195 121	E	101121	102121	103121
195 122	E	101122	102122	103122
195 123	E	101123	102123	103123
195 124	E	101124	102124	103124
195 125	E	101125	102125	103125
195 126	E	101126	102126	103126
195 127	E	101127	102127	103127
195 128	E	101128	102128	103128
195 129	E	101129	102129	103129
195 130	E	101130	102130	103130

2. DIESEL ELECTRIC UNITS

CLASS 201/202   PRESERVED "HASTINGS" UNIT   BR

DMBS–TSL–TSL–TSRB–TSL–DMBS.

Preserved unit made up from two Class 201 short-frame cars and three Class 202 long-frame cars. The "Hastings" units were made with narrow body-profiles for use on the section between Tonbridge and Battle which had tunnels of restricted loading gauge. These tunnels were converted to single track operation in the 1980s thus allowing standard loading gauge stock to be used. The set also contains a Class 411 EMU trailer (not Hastings line gauge) and a Class 422 EMU buffet car.

Construction: Steel.
Engine: One English Electric 4SRKT Mk. 2 of 450 kW (600 hp) at 850 rpm.
Main Generator: English Electric EE824.
Traction Motors: Two English Electric EE507 mounted on the inner bogie.
Bogies: SR Mk 4. (Former EMU TSL vehicles have Commonwealth bogies).
Couplers: Drophead buckeye.

Dimensions: 18.40 x 2.50 m (60000), 20.35 x 2.50 m (60116/118/529), 18.36 x 2.50 m (60501), 20.35 x 2.82 (69337), 20.30 x 2.82 (70262).
Gangways: Within unit only.   **Doors:** Manually operated slam.
Wheel arrangement: 2-Bo + 2-2 + 2-2 + 2-2- + 2-2- + Bo-2.
Brakes: Electro-pneumatic and automatic air.
Maximum Speed: 75 mph.   **Seating Layout:** 2+2 facing.
Multiple Working: Other ex-BR Southern Region DEMU vehicles.

60000. DMBS. Lot No. 30329 Eastleigh 1957. –/22. 55.0 t.
60116. DMBS. Lot No. 30395 Eastleigh 1957. –/31. 56.0 t.
60118. DMBS. Lot No. 30395 Eastleigh 1957. –/30. 56.0 t.
60501. TSL. Lot No. 30331 Eastleigh 1957. –/52 2T. 29.5 t.
60529. TSL. Lot No. 30397 Eastleigh 1957. –/60 2T. 30.5 t.
69337. TSRB (ex-Class 422 EMU). Lot No. 30805 York 1970. –/40. 35.0 t.
70262. TSL (ex-Class 411/5 EMU). Lot No. 30455 Eastleigh 1958. –/64 2T. 31.5 t.

201 001	**G**	HD	*HD*	SE	60116	60529	70262	69337	60501	60118
Spare	**G**	HD	*HD*	SE	60000					

Names:

60000	Hastings	60118	Tunbridge Wells
60116	Mountfield		

CLASS 220   VOYAGER   BOMBARDIER

DMS–MS–MS–DMF. All engines have been derated from 750 hp to 700 hp.

Construction: Steel.
Engine: Cummins QSK19 of 520 kW (700 hp) at 1800 rpm.
Transmission: Two Alstom Onix 800 three-phase traction motors of 275 kW.
Braking: Rheostatic and electro-pneumatic.
Bogies: Bombardier B5005.
Couplers: Dellner 12 at outer ends, bar within unit.
Dimensions: 23.85/23.00 (602xx) x 2.73 m.
Gangways: Within unit only.
Wheel Arrangement: 1A-A1 + 1A-A1 + 1A-A1 + 1A-A1.
Doors: Single-leaf swing plug.
Maximum Speed: 125 m.p.h.
Seating Layout: 1: 2+1 facing/unidirectional, 2: 2+2 mainly unidirectional.
Multiple Working: Within class and with Classes 221 and 222 (in an emergency). Also can be controlled from Class 57/3 locomotives.

DMS. Bombardier Bruges/Wakefield 2000–01. –/42 1TD 1W. 51.1 t.
MS(A). Bombardier Bruges/Wakefield 2000–01. –/66. 45.9 t.
MS(B). Bombardier Bruges/Wakefield 2000–01. –/66 1TD. 46.7 t.
DMF. Bombardier Bruges/Wakefield 2000–01. 26/– 1TD 1W. 50.9 t.

220 001	**XC**	BN	*XC*	CZ	60301	60701	60201	60401
220 002	**XC**	BN	*XC*	CZ	60302	60702	60202	60402
220 003	**XC**	BN	*XC*	CZ	60303	60703	60203	60403
220 004	**XC**	BN	*XC*	CZ	60304	60704	60204	60404
220 005	**XC**	BN	*XC*	CZ	60305	60705	60205	60405
220 006	**XC**	BN	*XC*	CZ	60306	60706	60206	60406
220 007	**XC**	BN	*XC*	CZ	60307	60707	60207	60407

220 008	XC	BN	CZ	60308	60708	60408
220 009	XC	BN	CZ	60309	60709	60409
220 010	XC	BN	CZ	60310	60710	60410
220 011	XC	BN	CZ	60311	60711	60411
220 012	XC	BN	CZ	60312	60712	60412
220 013	XC	BN	CZ	60313	60713	60413
220 014	XC	BN	CZ	60314	60714	60414
220 015	XC	BN	CZ	60315	60715	60415
220 016	XC	BN	CZ	60316	60716	60416
220 017	XC	BN	CZ	60317	60717	60417
220 018	XC	BN	CZ	60318	60718	60418
220 019	XC	BN	CZ	60319	60719	60419
220 020	XC	BN	CZ	60320	60720	60420
220 021	XC	BN	CZ	60321	60721	60421
220 022	XC	BN	CZ	60322	60722	60422
220 023	XC	BN	CZ	60323	60723	60423
220 024	XC	BN	CZ	60324	60724	60424
220 025	XC	BN	CZ	60325	60725	60425
220 026	XC	BN	CZ	60326	60726	60426
220 027	XC	BN	CZ	60327	60727	60427
220 028	XC	BN	CZ	60328	60728	60428
220 029	XC	BN	CZ	60329	60729	60429
220 030	XC	BN	CZ	60330	60730	60430
220 031	XC	BN	CZ	60331	60731	60431
220 032	XC	BN	CZ	60332	60732	60432
220 033	XC	BN	CZ	60333	60733	60433
220 034	XC	BN	CZ	60334	60734	60434

Construction: Steel.
Engine: Cummins QSK19 of 520 kW (700 hp) at 1800 rpm.
Transmission: Two Alstom Onix 800 three-phase traction motors of 275 kW.
Braking: Rheostatic and electro-pneumatic.
Bogies: Bombardier HVP.
Couplers: Dellner 12 at outer ends, bar within unit.
Dimensions: 23.67 x 2.73 m.
Gangways: Within unit only.
Wheel Arrangement: 1A-A1 + 1A-A1 + 1A-A1 (+ 1A-A1) + 1A-A1.
Doors: Single-leaf swing plug.
Maximum Speed: 125 mph.
Seating Layout: 1: 2+1 facing/unidirectional. 2: 2+2 mainly unidirectional.
Multiple Working: Within class and with Classes 220 and 222 (in an emergency). Also can be controlled from Class 57/3 locomotives.

* Virgin Trains units. MSRMB moved adjacent to the DMF. The seating in this vehicle (2+2 facing) can be used by First or Standard Class passengers depending on demand.

CLASS 221 SUPER VOYAGER BOMBARDIER

The * DMS–MS–MS–MSRMB–DMF (Virgin Trains units) or DMS–MS–(MS)–MS–DMF (CrossCountry units). Built as tilting units but tilt now isolated on CrossCountry sets. All engines have been derated from 750 hp to 700 hp.

Names (carried on MSRMB or DMS (222 003)):

222 001 THE ENTREPRENEUR EXPRESS
222 002 THE CUTLERS' COMPANY
222 003 TORNADO
222 004 CHILDREN'S HOSPITAL SHEFFIELD
222 006 THE CARBON CUTTER

222 007–023. 5-car units. DMF–MC–MSRMB–MS–DMS.

DMRF. Bombardier Bruges 2003–04. 22/– 1TD 1W. 52.8 t.
MC. Bombardier Bruges 2003–04. 28/22 1T. 48.6 t.
MSRMB. Bombardier Bruges 2003–04. –/62. 49.6 t.
MS. Bombardier Bruges 2004–05. –/68 1T. 47.0 t.
DMS. Bombardier Bruges 2003–04. –/40 1TD 1W. 51.0 t.

222 007	**ST**	E	*EM*	DY	60247	60442	60627	60567	60167
222 008	**ST**	E	*EM*	DY	60248	60918	60628	60545	60168
222 009	**ST**	E	*EM*	DY	60249	60919	60629	60557	60169
222 010	**ST**	E	*EM*	DY	60250	60920	60630	60546	60170
222 011	**ST**	E	*EM*	DY	60251	60921	60631	60531	60171
222 012	**ST**	E	*EM*	DY	60252	60922	60632	60532	60172
222 013	**ST**	E	*EM*	DY	60253	60923	60633	60533	60173
222 014	**ST**	E	*EM*	DY	60254	60924	60634	60534	60174
222 015	**ST**	E	*EM*	DY	60255	60925	60635	60535	60175
222 016	**ST**	E	*EM*	DY	60256	60926	60636	60536	60176
222 017	**ST**	E	*EM*	DY	60257	60927	60637	60537	60177
222 018	**ST**	E	*EM*	DY	60258	60928	60638	60444	60178
222 019	**ST**	E	*EM*	DY	60259	60929	60639	60547	60179
222 020	**ST**	E	*EM*	DY	60260	60930	60640	60543	60180
222 021	**ST**	E	*EM*	DY	60261	60931	60641	60552	60181
222 022	**ST**	E	*EM*	DY	60262	60932	60642	60542	60182
222 023	**ST**	E	*EM*	DY	60263	60933	60643	60541	60183

Names (carried on MSRMB or DMS):

222 008 Derby Etches Park
222 011 Sheffield City Battalion 1914–1918
222 015 175 YEARS OF DERBY'S RAILWAYS 1839–2014
222 017 LIONS CLUB INTERNATIONAL CENTENARY 1917–2017
222 022 INVEST IN NOTTINGHAM

222 101–104. 4-car former Hull Trains units. DMF–MC–MSRMB–DMS.

DMRF. Bombardier Bruges 2005. 22/– 1TD 1W. 52.8 t.
MC. Bombardier Bruges 2005. 11/46 1T. 47.1 t.
MSRMB. Bombardier Bruges 2005. –/62. 48.0 t.
DMS. Bombardier Bruges 2005. –/40 1TD 1W. 49.4 t.

222 101	**ST**	E	*EM*	DY	60271	60571	60191
222 102	**ST**	E	*EM*	DY	60272	60572	60192
222 103	**ST**	E	*EM*	DY	60273	60573	60193
222 104	**ST**	E	*EM*	DY	60274	60574	60194

▲ Southern-liveried 171 722 and 171 402 approach Norwood Junction with the 09.02 London Bridge–Uckfield on 19/06/17. **Brian Denton**

▼ Chiltern Railways-liveried 172 101 passes West Hampstead (LUL) with the 19.36 London Marylebone–Gerrards Cross on 05/07/16. **Robert Pritchard**

▲ Arriva Trains-liveried 175 113 is seen between Oxford Road and Piccadilly stations in Manchester with the 08.54 Llandudno Junction–Manchester Airport on 29/11/16. **Robert Pritchard**

▼ First Hull Trains Dynamic Lines-liveried 180 111 passes Claypole with the 15.48 London King's Cross–Hull on 10/08/17. **Paul Biggs**

▲ TransPennine Express liveried 185 138 leaves Lancaster with the 08.29 Manchester Airport–Windermere on 22/04/17. **Robert Pritchard**

▼ CrossCountry-liveried 220 008 arrives in the turnback siding just beyond Southampton Central station on 17/06/17. **Robert Pritchard**

▲ Virgin Trains-liveried 221 101 arrives at Milton Keynes Central on 26/02/17 with the 11.58 Lancaster–London Euston. **Robert Pritchard**

▼ East Midlands Trains-liveried 222 011 passes Cossington with the 07.04 Lincoln–London St Pancras on 19/08/17. **Paul Biggs**

▲ Vivarail-liveried D-Train prototype 230 001 is seen at Honeybourne running empty stock from Long Marston to Moreton-in-Marsh on 14/06/17.
Dave Gommersall

▼ Colas Rail Plasser & Theurer 08-4x4/4S-RT Tamper DR 73906 crosses Dent Head Viaduct with a 10.52 Garsdale–Guide Bridge Sidings on 03/05/17. **Paul Biggs**

▲ Volker Rail Matisa R 24 S Ballast Regulator DR 77802 is seen at Sutton Bonington with a 10.00 Melton Mowbray–Scunthorpe Frodingham on 06/02/15. **Paul Biggs**

▼ Brand new Network Rail Loram C44 Rail Grinding Train 99 70 9427 042-5 + 99 70 9427 043-3 + 99 70 9427 044-1 + 99 70 9427 045-8 ("DR 79401–404") is seen on test at Okehampton on the Dartmoor Railway on 10/05/17. **David Hunt**

▲ Network Rail Harsco Track Technologies Multi-purpose Stoneblower DR 80303 passes Virginia Water whilst running from Bristol Kingsland Road to Woking Up Yard on 19/06/17. **Robert Pritchard**

▼ Network Rail Plasser & Theurer NB-PW Ballast System Propulsion Machine DR 92264 is seen on the rear of a Taunton Fairwater Yard–Westbury train at Cogload Junction on 03/08/17. **Tony Christie**

▲ Network Rail Robel Type 69.70 Mobile Maintenance Train 99 70 9481 005-5 + 99 70 9559 005-2 + 99 70 9580 005-5 ("DR 97505/605/805") is seen at Sheffield on 05/07/16. **Andy Barclay**

▼ Track Assessment unit 999600+999601 ("950 001") (based on the BREL Class 150/1 design) passes Trowell heading for Derby RTC on 14/03/17. **Steve Donald**

CLASS 230 D-TRAIN METRO-CAMMELL/VIVARAIL

The Class 230 D-Train is a prototype 3-car DEMU rebuilt from former London Underground D78 Stock by Vivarail at Long Marston. The D-Train uses the bodyshells, bogies and electric traction motors of D78 Stock. Instead of being powered by electricity the motors are instead powered by new underfloor-mounted diesel engines – two per driving car. Modern IGBT electronic controls replace the previous mechanical camshaft controllers, incorporating the latest automotive stop-start technology and dynamic braking.

The prototype was due to commence trial operation on the Coventry–Nuneaton line in 2017, but following fire damage sustained during test running in December 2016 this trial has been cancelled. Full interior details are awaited but include a mix of seating layouts and a universal access toilet in the TSO.

Vivarail has acquired more than 200 redundant D78 Stock vehicles that are stored at Long Marston and it is hoped that orders for further conversions will be forthcoming.

Construction: Aluminium.
Engine: 2 x Ford Duratorq 3.2 litre engines of 150 kW (200 hp).
Control System: IGBT Inverter.   **Braking:** Rheostatic & Dynamic.
Bogies: Bombardier Flex1000 flexible-frame.
Dimensions: 18.37/18.12 x 2.85 m.
Couplers: LUL automatic wedgelock. **Gangways:** Within unit only.
Wheel Arrangement: Bo-Bo + 2-2 + Bo-Bo.
Doors: Sliding.   **Maximum Speed:** 60 mph.
Seating Layout: Longitudinal or 2+2 facing.
Multiple Working: Within class.

Rebuilt from former London Underground D78 Stock:
300001 rebuilt from Driving Motor 7058; redesignated DMSO(A).
300101 rebuilt from Driving Motor 7511; redesignated DMSO(B).
300201 rebuilt from Trailer 17128; redesignated TSO.

DMSO(A). Metro-Cammell Birmingham 1979–83. 28 t.
TSO. Metro-Cammell Birmingham 1979–83. 20 t.
DMSO(B). Metro-Cammell Birmingham 1979–83. 28 t.

230 001	**VI** VI		LM	300001	300201	300101

3. DMUS AWAITING DISPOSAL

The list below comprises vehicles which are stored awaiting disposal.

Non-Standard Livery: 121 020 All over Chiltern blue with a silver stripe.

Class 121

121 020	**0**	CR	AL	55020
121 034	**G**	CR	AL	55034

4. ON-TRACK MACHINES

These machines are used for maintaining, renewing and enhancing the infrastructure of the national railway network. With the exception of snowploughs all can be self-propelled, controlled either from a cab mounted on the machine or remotely. They are permitted to operate either under their own power or in train formations throughout the network both within and outside engineering possessions. Machines only permitted to be used within engineering possessions, referred to as On-Track Plant, are not included. Also not included are wagons included in OTM consists.

For each machine its Network Rail registered number, owner or responsible custodian and type is given, plus its name if carried. In addition, for snow clearance equipment the berthing location is given. Actual operation of each machine is undertaken by either the owner/responsible custodian or a contracted responsible custodian.

Machines were numbered by British Rail with either six-digit wagon series numbers or in the CEPS (Civil Engineers Plant System) series with five prefixed digits. Recently delivered machines have been numbered in the EVN series. Machines may also carry additional identifying numbers which are shown as "xxxx". Machines are listed here in CEPS/wagon series order. Those with EVN numbers are included where they would have been if allocated CEPS numbers.

(S) after the registered number designates a machine that is currently stored (the storage location of each is given at the end of this section).

DYNAMIC TRACK STABILISERS

| DR 72211 | BB | Plasser & Theurer DGS 62-N |
| DR 72213 | BB | Plasser & Theurer DGS 62-N |

TAMPERS

DR 73108	CS	Plasser & Theurer 09-32-RT	Tiger
DR 73109	SK	Plasser & Theurer 09-3X-RT	
DR 73110	SK	Plasser & Theurer 09-3X-RT	PETER WHITE
DR 73111	NR	Plasser & Theurer 09-3X-Dynamic	
DR 73113	NR	Plasser & Theurer 09-3X-Dynamic	
DR 73114	NR	Plasser & Theurer 09-3X-Dynamic	Ron Henderson
DR 73115	NR	Plasser & Theurer 09-3X-Dynamic	
DR 73116	NR	Plasser & Theurer 09-3X Dynamic	
DR 73117	NR	Plasser & Theurer 09-3X Dynamic	
DR 73118	NR	Plasser & Theurer 09-3X Dynamic	

99 70 9123 120-6	NR	Plasser & Theurer 09-3X Dynamic "DR 73120"
99 70 9123 121-4	NR	Plasser & Theurer 09-2X Dynamic "DR 73121"
99 70 9123 122-2	NR	Plasser & Theurer 09-2X Dynamic "DR 73122"

| DR 73803 | SK | Plasser & Theurer 08-32U-RT | Alexander Graham Bel |

DR 73804	SK	Plasser & Theurer 08-32U-RT	James Watt
DR 73805	CS	Plasser & Theurer 08-16/32U-RT	
DR 73806	CS	Plasser & Theurer 08-16/32U-RT	Karine
DR 73904	SK	Plasser & Theurer 08-4x4/4S-RT	Thomas Telford
DR 73905	CS	Plasser & Theurer 08-4x4/4S-RT	
DR 73906	CS	Plasser & Theurer 08-4x4/4S-RT	Panther
DR 73907	CS	Plasser & Theurer 08-4x4/4S-RT	
DR 73908	CS	Plasser & Theurer 08-4x4/4S-RT	
DR 73909	CS	Plasser & Theurer 08-4x4/4S-RT	Saturn
DR 73910	CS	Plasser & Theurer 08-4x4/4S-RT	Jupiter
DR 73911	CS	Plasser & Theurer 08-16/4x4C-RT	Puma
DR 73912	CS	Plasser & Theurer 08-16/4x4C-RT	Lynx
DR 73913	CS	Plasser & Theurer 08-12/4x4C-RT	
DR 73914	SK	Plasser & Theurer 08-4x4/4S-RT	Robert McAlpine
DR 73915	SK	Plasser & Theurer 08-16/4x4C-RT	William Arrol
DR 73916	SK	Plasser & Theurer 08-16/4x4C-RT	First Engineering
DR 73917	BB	Plasser & Theurer 08-4x4/4S-RT	
DR 73918	BB	Plasser & Theurer 08-4x4/4S-RT	
DR 73919	CS	Plasser & Theurer 08-16/4x4C100-RT	
DR 73920	CS	Plasser & Theurer 08-16/4x4C80-RT	
DR 73921	CS	Plasser & Theurer 08-16/4x4C80-RT	
DR 73922	CS	Plasser & Theurer 08-16/4x4C80-RT	John Snowdon
DR 73923	CS	Plasser & Theurer 08-4x4/4S-RT	Mercury
DR 73924	CS	Plasser & Theurer 08-16/4x4C100-RT	
DR 73925	CS	Plasser & Theurer 08-16/4x4C100-RT	Europa
DR 73926	BB	Plasser & Theurer 08-16/4x4C100-RT	Stephen Keith Blanchard
DR 73927	BB	Plasser & Theurer 08-16/4x4C100-RT	
DR 73928	BB	Plasser & Theurer 08-16/4x4C100-RT	
DR 73929	CS	Plasser & Theurer 08-4x4/4S-RT	
DR 73930	CS	Plasser & Theurer 08-4x4/4S-RT	
DR 73931	CS	Plasser & Theurer 08-16/4x4C100-RT	
DR 73932	SK	Plasser & Theurer 08-4x4/4S-RT	
DR 73933	SK	Plasser & Theurer 08-16/4x4/C100-RT	
DR 73934	SK	Plasser & Theurer 08-16/4x4/C100-RT	
DR 73935	CS	Plasser & Theurer 08-4x4/4S-RT	
DR 73936	CS	Plasser & Theurer 08-4x4/4S-RT	
DR 73937	BB	Plasser & Theurer 08-16/4x4C100-RT	
DR 73938	BB	Plasser & Theurer 08-16/4x4C100-RT	
DR 73939	BB	Plasser & Theurer 08-16/4x4C100-RT	Pat Best
DR 73940	SK	Plasser & Theurer 08-4x4/4S-RT	
DR 73941	SK	Plasser & Theurer 08-4x4/4S-RT	
DR 73942	CS	Plasser & Theurer 08-4x4/4S-RT	
DR 73943	BB	Plasser & Theurer 08-16/4x4C100-RT	
DR 73944	BB	Plasser & Theurer 08-16/4x4C100-RT	
DR 73945	BB	Plasser & Theurer 08-16/4x4C100-RT	
DR 73946	VO	Plasser & Theurer Euromat 08-4x4/4S	
DR 73947	CS	Plasser & Theurer 08-4x4/4S-RT	
DR 73948	CS	Plasser & Theurer 08-4x4/4S-RT	
9 70 9128 001-3	SK	Plasser & Theurer Unimat 09-4x4/4S Dynamic "928001"	
9 70 9128 002-1	SK	Plasser & Theurer Unimat 09-4x4/4S Dynamic "DR 74002"	

DR 75301	VO	Matisa B 45 UE	
DR 75302	VO	Matisa B 45 UE	Gary Wright
DR 75303	VO	Matisa B 45 UE	
DR 75401	VO	Matisa B 41 UE	
DR 75402	VO	Matisa B 41 UE	
DR 75403 (S)	VO	Matisa B 41 UE	
DR 75404	VO	Matisa B 41 UE	
DR 75405	VO	Matisa B 41 UE	
DR 75406	CS	Matisa B 41 UE	Eric Machell
DR 75407	CS	Matisa B 41 UE	
DR 75408	BB	Matisa B 41 UE	
DR 75409	BB	Matisa B 41 UE	
DR 75410	BB	Matisa B 41 UE	
DR 75411	BB	Matisa B 41 UE	
DR 75501	BB	Matisa B 66 UC	
DR 75502	BB	Matisa B 66 UC	

BALLAST CLEANERS

DR 76323	NR	Plasser & Theurer RM95-RT	
DR 76324	NR	Plasser & Theurer RM95-RT	
DR 76501	NR	Plasser & Theurer RM-900-RT	
DR 76502	NR	Plasser & Theurer RM-900-RT	
DR 76503	NR	Plasser & Theurer RM-900-RT	
99 70 9314 504-0	NR	Plasser & Theurer RM-900	"DR 76504"

VACUUM PREPARATION MACHINES

DR 76701	NR	Plasser & Theurer VM80-NR
DR 76702	NR	Plasser & Theurer VM80-NR
DR 76703	NR	Plasser & Theurer VM80-NR
DR 76710 (S)	NR	Plasser & Theurer VM80-TRS
DR 76711 (S)	NR	Plasser & Theurer VM80-TRS

RAIL VACUUM MACHINES

99 70 9515 001-4	RC	Railcare 16000-480-UK RailVac OTM
99 70 9515 002-2	RC	Railcare 16000-480-UK RailVac OTM
99 70 9515 003-0	RC	Railcare 16000-480-UK RailVac OTM
99 70 9515 004-8	RC	Railcare 16000-480-UK RailVac OTM
99 70 9515 005-5	RC	Railcare 16000-480-UK RailVac OTM
99 70 9515 006-3	RC	Railcare 16000-480-UK RailVac OTM

BALLAST FEEDER MACHINE

| 99 70 9552 020-5 | RC | Railcare Ballast Feeder UK |

BALLAST TRANSFER MACHINES

| DR 76750 | NR | Matisa D75 | (works with DR 78802/DR 78812/ DR 78822/DR 78832) |
| DR 76751 | NR | Matisa D75 | (works with DR 78801/DR 78811/ DR 78821/DR 78831) |

CONSOLIDATION MACHINES

| DR 76801 | NR | Plasser & Theurer 09-CM-NR |
| 99 70 9320 802-0 | NR | Plasser & Theurer 09-2X-CM "DR 76802" |

FINISHING MACHINES & BALLAST REGULATORS

DR 77001	SK	Plasser & Theurer AFM 2000-RT Finishing Machine
DR 77002	SK	Plasser & Theurer AFM 2000-RT Finishing Machine
99 70 9125 010-7	NR	Plasser & Theurer USP 6000 Regulator "DR 77010"

DR 77315 (S)	BB	Plasser & Theurer USP 5000C Regulator
DR 77316 (S)	BB	Plasser & Theurer USP 5000C Regulator
DR 77322	BB	Plasser & Theurer USP 5000C Regulator
DR 77327	CS	Plasser & Theurer USP 5000C Regulator
DR 77336 (S)	BB	Plasser & Theurer USP 5000C Regulator
DR 77801	VO	Matisa R 24 S Regulator
DR 77802	VO	Matisa R 24 S Regulator
DR 77901	CS	Plasser & Theurer USP 5000-RT Regulator
DR 77903	NR	Plasser & Theurer USP 5000-RT Regulator
DR 77904	NR	Plasser & Theurer USP 5000-RT Regulator
DR 77905	NR	Plasser & Theurer USP 5000-RT Regulator
DR 77906	NR	Plasser & Theurer USP 5000-RT Regulator
DR 77907	NR	Plasser & Theurer USP 5000-RT Regulator
DR 77908	SK	Plasser & Theurer USP 5000-RT Regulator
99 70 9125 909-0	NR	Plasser & Theurer USP 5000 Regulator "DR 77909"

TWIN JIB TRACK RELAYERS

DRP 78213	VO	Plasser & Theurer Self-Propelled Heavy Duty
DRP 78215	SK	Plasser & Theurer Self-Propelled Heavy Duty
DRP 78216	BB	Plasser & Theurer Self-Propelled Heavy Duty
DRP 78217 (S)	SK	Plasser & Theurer Self-Propelled Heavy Duty
DRP 78218 (S)	BB	Plasser & Theurer Self-Propelled Heavy Duty
DRP 78219	SK	Plasser & Theurer Self-Propelled Heavy Duty
DRP 78221	BB	Plasser & Theurer Self-Propelled Heavy Duty
DRP 78222	BB	Plasser & Theurer Self-Propelled Heavy Duty
DRP 78223 (S)	BB	Plasser & Theurer Self-Propelled Heavy Duty
DRP 78224 (S)	BB	Plasser & Theurer Self-Propelled Heavy Duty
DRC 78226	CS	Cowans Sheldon Self-Propelled Heavy Duty
DRC 78229	NR	Cowans Sheldon Self-Propelled Heavy Duty

DRC 78231	NR	Cowans Sheldon Self-Propelled Heavy Duty
DRC 78234	NR	Cowans Sheldon Self-Propelled Heavy Duty
DRC 78235	CS	Cowans Sheldon Self-Propelled Heavy Duty
DRC 78237 (S)	CS	Cowans Sheldon Self-Propelled Heavy Duty

NEW TRACK CONSTRUCTION
TRAIN PROPULSION MACHINES

| DR 78701 | BB | Harsco Track Technologies NTC-PW |
| DR 78702 | BB | Harsco Track Technologies NTC-PW |

TRACK RENEWAL MACHINES

Matisa P95 Track Renewals Trains
DR 78801+DR 78811+DR 78821+DR 78831 NR *(works with DR 76751)*
DR 78802+DR 78812+DR 78822+DR 78832 NR *(works with DR 76750)*

RAIL GRINDING TRAINS

Loram SPML 15
DR 79200A + DR 79200B + DR 79200C   NR

Loram SPML 17
DR 79201A + DR 79201B   NR

Speno RPS-32
DR 79221 + DR 79222 + DR 79223 + DR 79224 + DR 79225 + DR 79226   SI

Loram C21
DR 79231 + DR 79232 + DR 79233 + DR 79234 + DR 79235 + DR 79236 + DR 79237   NR
DR 79241 + DR 79242 + DR 79243 + DR 79244 + DR 79245 + DR 79246 + DR 79247   NR
DR 79251 + DR 79252 + DR 79253 + DR 79254 + DR 79255 + DR 79256 + DR 79257   NR

Names: DR 79241/247   Roger South *(one plate on opposite sides of each)*
                   DR 79251/257   Martin Elwood

Harsco Track Technologies RGH20C
DR 79261 + DR 79271	NR	
DR 79262 + DR 79272	NR	
DR 79263 + DR 79273	NR	
DR 79264 + DR 79274	NR	
DR 79265 (S)	NR	*spare vehicle*
DR 79267 + DR 79277	NR	

Loram C44
99 70 9427 038-3 + 99 70 9427 039-1 + 99 70 9427 040-9 + 99 70 9427 041-
                              NR   "DR 79301/302/303/304"
99 70 9427 042-5 + 99 70 9427 043-3 + 99 70 9427 044-1 + 99 70 9427 045-
                              NR   "DR 79401/402/403/404"
99 70 9427 046-6 + 99 70 9427 047-4 + 99 70 9427 048-2 + 99 70 9427 049-
99 70 9427 050-8 + 99 70 9427 051-6 + 99 70 9427 052-4
                              NR   "DR 79501/502/503/504/505/506/507"

STONEBLOWERS

DR 80200 (S)	NR	Pandrol Jackson Plain Line	
DR 80201	NR	Pandrol Jackson Plain Line	
DR 80202 (S)	NR	Pandrol Jackson Plain Line	
DR 80203 (S)	NR	Pandrol Jackson Plain Line	
DR 80204 (S)	NR	Pandrol Jackson Plain Line	
DR 80205	NR	Pandrol Jackson Plain Line	
DR 80206	NR	Pandrol Jackson Plain Line	
DR 80207 (S)	NR	Pandrol Jackson Plain Line	
DR 80208	NR	Pandrol Jackson Plain Line	
DR 80209	NR	Pandrol Jackson Plain Line	
DR 80210	NR	Pandrol Jackson Plain Line	
DR 80211	NR	Pandrol Jackson Plain Line	
DR 80212 (S)	NR	Pandrol Jackson Plain Line	
DR 80213	NR	Harsco Track Technologies Plain Line	
DR 80214	NR	Harsco Track Technologies Plain Line	
DR 80215	NR	Harsco Track Technologies Plain Line	
DR 80216	NR	Harsco Track Technologies Plain Line	
DR 80217	NR	Harsco Track Technologies Plain Line	
DR 80301	NR	Harsco Track Technologies Multi-purpose	Stephen Cornish
DR 80302	NR	Harsco Track Technologies Multi-purpose	
DR 80303	NR	Harsco Track Technologies Multi-purpose	

CRANES

DRP 81505	BB	Plasser & Theurer 12 tonne Heavy Duty Diesel Hydraulic
DRP 81507 (S)	BB	Plasser & Theurer 12 tonne Heavy Duty Diesel Hydraulic
DRP 81508	BB	Plasser & Theurer 12 tonne Heavy Duty Diesel Hydraulic
DRP 81511 (S)	BB	Plasser & Theurer 12 tonne Heavy Duty Diesel Hydraulic
DRP 81513	BB	Plasser & Theurer 12 tonne Heavy Duty Diesel Hydraulic
DRP 81517	BB	Plasser & Theurer 12 tonne Heavy Duty Diesel Hydraulic
DRP 81519 (S)	BB	Plasser & Theurer 12 tonne Heavy Duty Diesel Hydraulic
DRP 81522	BB	Plasser & Theurer 12 tonne Heavy Duty Diesel Hydraulic
DRP 81525	BB	Plasser & Theurer 12 tonne Heavy Duty Diesel Hydraulic
DRP 81532	BB	Plasser & Theurer 12 tonne Heavy Duty Diesel Hydraulic
DRK 81601	VO	Kirow KRC 810UK 100 tonne Heavy Duty Diesel Hydraulic
DRK 81602	BB	Kirow KRC 810UK 100 tonne Heavy Duty Diesel Hydraulic
DRK 81611	BB	Kirow KRC 1200UK 125 tonne Heavy Duty Diesel Hydraulic
DRK 81612	CS	Kirow KRC 1200UK 125 tonne Heavy Duty Diesel Hydraulic
DRK 81613	VO	Kirow KRC 1200UK 125 tonne Heavy Duty Diesel Hydraulic
DRK 81621	VO	Kirow KRC 250UK 25 tonne Diesel Hydraulic
DRK 81622	VO	Kirow KRC 250UK 25 tonne Diesel Hydraulic
DRK 81623	SK	Kirow KRC 250UK 25 tonne Diesel Hydraulic
DRK 81624	SK	Kirow KRC 250UK 25 tonne Diesel Hydraulic
DRK 81625	SK	Kirow KRC 250UK 25 tonne Diesel Hydraulic
99 70 9319 012-9	SK	Kirow KRC 250S 25 tonne Diesel Hydraulic "DRK 81626"
99 70 9319 013-7	NR	Kirow KRC 1200UK 125 tonne Heavy Duty Diesel Hydraulic

Names:

DRK 81601   Nigel Chester | DRK 81611   Malcolm L. Pearce

LONG WELDED RAIL TRAIN PROPULSION MACHINES

DR 89005   NR   Cowans Boyd PW
DR 89007   NR   Cowans Boyd PW
DR 89008   NR   Cowans Boyd PW

BALLAST SYSTEM PROPULSION MACHINES

DR 92263 (S)   NR   Plasser & Theurer MFS-PW
DR 92264   NR   Plasser & Theurer NB-PW
DR 92285   NR   Plasser & Theurer PW-RT
DR 92286   NR   Plasser & Theurer NPW-RT
DR 92331   NR   Plasser & Theurer PW-RT
DR 92332   NR   Plasser & Theurer NPW-RT
DR 92431   NR   Plasser & Theurer PW-RT
DR 92432   NR   Plasser & Theurer NPW-RT
99 70 9310 477-3 NR   Plasser & Theurer PW   "DR 92477"
99 70 9310 478-1 NR   Plasser & Theurer NPW   "DR 92478"

BREAKDOWN CRANES

ADRC 96710 (S)   NR   Cowans Sheldon 75 tonne Diesel Hydraulic
ADRC 96713   NR   Cowans Sheldon 75 tonne Diesel Hydraulic
ADRC 96714 (S)   NR   Cowans Sheldon 75 tonne Diesel Hydraulic
ADRC 96715   NR   Cowans Sheldon 75 tonne Diesel Hydraulic

GENERAL PURPOSE VEHICLES

DR 97001   H1   Eiv de Brieve DU94BA TRAMM with Crane "DU 94 B 001 URS"

DR 97011   H1   Windhoff MPV (Modular)
DR 97012   H1   Windhoff MPV (Modular)
DR 97013   H1   Windhoff MPV (Modular)
DR 97014   H1   Windhoff MPV (Modular)

DR 98215A + DR 98215B   BB   Plasser & Theurer GP-TRAMM with Trailer
DR 98216A + DR 98216B   BB   Plasser & Theurer GP-TRAMM with Trailer
DR 98217A + DR 98217B   BB   Plasser & Theurer GP-TRAMM with Trailer
DR 98218A + DR 98218B   BB   Plasser & Theurer GP-TRAMM with Trailer
DR 98219A + DR 98219B   BB   Plasser & Theurer GP-TRAMM with Trailer
DR 98220A + DR 98220B   BB   Plasser & Theurer GP-TRAMM with Trailer

DR 98305 (S) NR   Geismar GP-TRAMM VMT 860 PL/UM
DR 98306 (S) NR   Geismar GP-TRAMM VMT 860 PL/UM

DR 98307A + DR 98307B (S) CS   Geismar GP-TRAMM VMT 860 PL/UM with Trailer
DR 98308A + DR 98308B (S) CS   Geismar GP-TRAMM VMT 860 PL/UM with Trailer

DR 98901 + DR 98951	NR	Windhoff MPV Master & Slave
DR 98902 + DR 98952	NR	Windhoff MPV Master & Slave
DR 98903 + DR 98953	NR	Windhoff MPV Master & Slave
DR 98904 + DR 98954	NR	Windhoff MPV Master & Slave
DR 98905 + DR 98955	NR	Windhoff MPV Master & Slave
DR 98906 + DR 98956	NR	Windhoff MPV Master & Slave
DR 98907 + DR 98957	NR	Windhoff MPV Master & Slave
DR 98908 + DR 98958	NR	Windhoff MPV Master & Slave
DR 98909 + DR 98959	NR	Windhoff MPV Master & Slave
DR 98910 + DR 98960	NR	Windhoff MPV Master & Slave
DR 98911 + DR 98961	NR	Windhoff MPV Master & Slave
DR 98912 + DR 98962	NR	Windhoff MPV Master & Slave
DR 98913 + DR 98963	NR	Windhoff MPV Master & Slave
DR 98914 + DR 98964	NR	Windhoff MPV Master & Slave
DR 98915 + DR 98965	NR	Windhoff MPV Master & Slave
DR 98916 + DR 98966	NR	Windhoff MPV Master & Slave
DR 98917 + DR 98967	NR	Windhoff MPV Master & Slave
DR 98918 + DR 98968	NR	Windhoff MPV Master & Slave
DR 98919 + DR 98969	NR	Windhoff MPV Master & Slave
DR 98920 + DR 98970	NR	Windhoff MPV Master & Slave
DR 98921 + DR 98971	NR	Windhoff MPV Master & Slave
DR 98922 + DR 98972	NR	Windhoff MPV Master & Slave
DR 98923 + DR 98973	NR	Windhoff MPV Master & Slave
DR 98924 + DR 98974	NR	Windhoff MPV Master & Slave
DR 98925 + DR 98975	NR	Windhoff MPV Master & Slave
DR 98926 + DR 98976	NR	Windhoff MPV Master & Powered Slave
DR 98927 + DR 98977	NR	Windhoff MPV Master & Powered Slave
DR 98928 + DR 98978	NR	Windhoff MPV Master & Powered Slave
DR 98929 + DR 98979	NR	Windhoff MPV Master & Powered Slave
DR 98930 + DR 98980	NR	Windhoff MPV Master & Powered Slave
DR 98931 + DR 98981	NR	Windhoff MPV Master & Powered Slave
DR 98932 + DR 98982	NR	Windhoff MPV Master & Powered Slave

Names:

DR 97012	Geoff Bell	DR 98923+DR 98973	Chris Lemon
DR 98914+DR 98964	Dick Preston	DR 98926+DR 98976	John Denyer
DR 98915+DR 98965	Nigel Cummins		

MOBILE MAINTENANCE TRAINS

Robel Type 69.70 Mobile Maintenance System

99 70 9481 001-4 + 99 70 9559 001-1 + 99 70 9580 001-4	NR	"DR 97501/601/801"
99 70 9481 002-2 + 99 70 9559 002-9 + 99 70 9580 002-2	NR	"DR 97502/602/802"
99 70 9481 003-0 + 99 70 9559 003-7 + 99 70 9580 003-0	NR	"DR 97503/603/803"
99 70 9481 004-8 + 99 70 9559 004-5 + 99 70 9580 004-8	NR	"DR 97504/604/804"
99 70 9481 005-5 + 99 70 9559 005-2 + 99 70 9580 005-5	NR	"DR 97505/605/805"
99 70 9481 006-3 + 99 70 9559 006-0 + 99 70 9580 006-3	NR	"DR 97506/606/806"
99 70 9481 007-1 + 99 70 9559 007-8 + 99 70 9580 007-1	NR	"DR 97507/607/807"
99 70 9481 008-9 + 99 70 9559 008-6 + 99 70 9580 008-9	NR	"DR 97508/608/808"

ELECTRIFICATION VEHICLES

DR 98001	NR	Windhoff MPV with Piling Equipment
DR 98002	NR	Windhoff MPV with Overhead Line Renewal Equipment
DR 98003	NR	Windhoff MPV with Overhead Line Renewal Equipment
DR 98004	NR	Windhoff MPV with Overhead Line Renewal Equipment
DR 98005	NR	Windhoff MPV with Overhead Line Renewal Equipment
DR 98006	NR	Windhoff MPV with Overhead Line Renewal Equipment
DR 98007	NR	Windhoff MPV with Piling Equipment
DR 98009	NR	Windhoff MPV with Overhead Line Renewal Equipment
DR 98010	NR	Windhoff MPV with Overhead Line Renewal Equipment
DR 98011	NR	Windhoff MPV with Overhead Line Renewal Equipment
DR 98012	NR	Windhoff MPV with Overhead Line Renewal Equipment
DR 98013	NR	Windhoff MPV with Overhead Line Renewal Equipment
DR 98014	NR	Windhoff MPV with Overhead Line Renewal Equipment

99 70 9131 001-8	NR	Windhoff MPV with Piling Equipment	"DR 76901"
99 70 9131 003-4	NR	Windhoff MPV with Piling Equipment	"DR 76903"
99 70 9131 005-9	NR	Windhoff MPV with Piling Equipment	"DR 76905"
99 70 9131 006-7	NR	Windhoff MPV with Concrete Equipment	"DR 76906"
99 70 9131 010-9	NR	Windhoff MPV with Concrete Equipment	"DR 76910"
99 70 9131 011-7	NR	Windhoff MPV with Structure Equipment	"DR 76911"
99 70 9131 013-3	NR	Windhoff MPV with Structure Equipment	"DR 76913"
99 70 9131 014-1	NR	Windhoff MPV with Overhead Line Equipment	"DR 76914"
99 70 9131 015-8	NR	Windhoff MPV with Overhead Line Equipment	"DR 76915"
99 70 9131 018-2	NR	Windhoff MPV with Overhead Line Equipment	"DR 76918"
99 70 9131 020-8	NR	Windhoff MPV with Overhead Line Equipment	"DR 76920"
99 70 9131 021-6	NR	Windhoff MPV with Overhead Line Equipment	"DR 76921"
99 70 9131 022-4	NR	Windhoff MPV with Final Works Equipment	"DR 76922"
99 70 9131 023-2	NR	Windhoff MPV with Final Works Equipment	"DR 76923"

99 70 9231 001-7	AM	SVI RT250 with crane & access platform
99 70 9231 004-1	AM	SVI PT500 with wire manipulator & access platform
99 70 9231 005-8	AM	SVI RSM9 with access platform
99 70 9231 006-6	AM	SVI RSM9 with access platform
99 70 9231 007-4	AM	APV250 with access platform

Names:

DR 98003	ANTHONY WRIGHTON 1944–2011
DR 98004	PHILIP CATTRELL 1961–2011
DR 98006	JASON MCDONNELL 1970–2016
DR 98009	MELVYN SMITH 1953–2011
DR 98010	BENJAMIN GAUTREY 1992–2011
DR 98012	TERENCE HAND 1962–2016
DR 98013	DAVID WOOD 1951–2015
DR 98014	WAYNE IMLACH 1955–2015
99 70 9131 001-8	BRUNEL
99 70 9131 023-2	GAVIN ROBERTS

INFRASTRUCTURE MONITORING VEHICLES

Note: "950 001" is a purpose-built Track Assessment Unit based on the BREL Class 150/1 design.

DR 98008	NR	Windhoff MPV Twin-cab with surveying equipment
999600 + 999601	NR	BREL York Track Assessment Unit "950 001"
999800 (S)	NR	Plasser & Theurer EM-SAT 100/RT Track Survey Car
999801 (S)	NR	Plasser & Theurer EM-SAT 100/RT Track Survey Car

Name:

999800   Richard Spoors

SNOWPLOUGHS

ADB 965203	NR	Independent Drift Plough	Carlisle Kingmoor Depot
ADB 965206	NR	Independent Drift Plough	York Leeman Road Sidings
ADB 965208	NR	Independent Drift Plough	Motherwell Depot
ADB 965209	NR	Independent Drift Plough	Taunton Fairwater Yard
ADB 965210	NR	Independent Drift Plough	Tonbridge West Yard
ADB 965211	NR	Independent Drift Plough	March Depot
ADB 965217	NR	Independent Drift Plough	Motherwell Depot
ADB 965219	NR	Independent Drift Plough	Motherwell Depot
ADB 965223	NR	Independent Drift Plough	Cardiff Canton Depot
ADB 965224	NR	Independent Drift Plough	Carlisle Kingmoor Depot
ADB 965230	NR	Independent Drift Plough	Carlisle Kingmoor Depot
ADB 965231	NR	Independent Drift Plough	Taunton Tamper Sidings
ADB 965234	NR	Independent Drift Plough	Inverness Millburn Yard
ADB 965235	NR	Independent Drift Plough	Cardiff Taff Vale Sidings
ADB 965236	NR	Independent Drift Plough	Tonbridge West Yard
ADB 965237	NR	Independent Drift Plough	March Depot
ADB 965240	NR	Independent Drift Plough	Motherwell Depot
ADB 965241	NR	Independent Drift Plough	York Yard North Sidings
ADB 965242	NR	Independent Drift Plough	Carlisle Kingmoor Depot
ADB 965243	NR	Independent Drift Plough	Inverness Millburn Yard
ADB 965576	NR	Beilhack Type PB600 Plough	Doncaster West Yard
ADB 965577	NR	Beilhack Type PB600 Plough	Doncaster West Yard
ADB 965578	NR	Beilhack Type PB600 Plough	Carlisle Kingmoor Yard
ADB 965579	NR	Beilhack Type PB600 Plough	Carlisle Kingmoor Yard
ADB 965580	NR	Beilhack Type PB600 Plough	Crewe Gresty Bridge Depot
ADB 965581	NR	Beilhack Type PB600 Plough	Crewe Gresty Bridge Depot
ADB 966098	NR	Beilhack Type PB600 Plough	Doncaster West Yard
ADB 966099	NR	Beilhack Type PB600 Plough	Doncaster West Yard

SNOWBLOWERS

ADB 968500	NR	Beilhack Self-Propelled Rotary	Edinburgh Slateford Depot
ADB 968501	NR	Beilhack Self-Propelled Rotary	Edinburgh Slateford Depot

ON-TRACK MACHINES AWAITING DISPOSAL

Tampers
DR 73105	Plasser & Theurer 09-32 CSM	Cardiff Canton Depot
DR 75201	Plasser & Theurer 08-275 S&C	Hither Green Depot
DR 75202	Plasser & Theurer 08-275 S&C	Hither Green Depot

Twin Jib track relayer
DRB 78123 British Hoist & Crane Non-Self-Propelled   Polmadie DHS

LOCATIONS OF STORED ON-TRACK MACHINES

The locations of machines shown above as stored (S) are shown here.

DR 75403	Frodingham OTM Depot	DR 80207	Eastleigh Works
DR 76710	Crewe Gresty Lane Sidings	DR 80212	Eastleigh Works
DR 76711	Taunton Fairwater Yard	DRP 81507	Ashford OTM Depot
DR 77315	Ashford OTM Depot	DRP 81511	Ashford OTM Depot
DR 77316	Ashford OTM Depot	DRP 81519	Woking OTM Depot
DR 77336	Hither Green Depot	DR 92263	Tyne Yard
DRP 78217	Glasgow Rutherglen Depot	ARDC 96710	Wigan Springs Branch Depot
DRP 78218	Ashford OTM Depot	ARDC 96714	Wigan Springs Branch Depot
DRP 78223	Ashford OTM Depot	DR 98305	Eastleigh Works
DRP 78224	Hither Green Depot	DR 98306	Eastleigh Works
DRC 78237	Rugby Depot	DR 98307A	Darley Dale
DR 79265	York Holgate Works	DR 98307B	Baglan Bay Yard
DR 80200	Thuxton	DR 98308A+	
DR 80202	Eastleigh Works	DR 98308B	Barry Rail Centre
DR 80203	Eastleigh Works	999800	Eastleigh Works
DR 80204	Thuxton	999801	Eastleigh Works

6. CODES

6.1. LIVERY CODES

1	"One" (metallic grey with a broad black bodyside stripe. White National Express/Greater Anglia "interim" stripe as branding).
AL	Advertising/promotional livery (see class heading for details).
AN	Anglia Railways Class 170s (white & turquoise with blue vignette).
AV	Arriva Trains (turquoise blue with white doors and a cream "swish").
AW	Arriva Trains Wales/Welsh Government sponsored dark & light blue.
CL	Chiltern Railways Mainline Class 168 (white & silver).
CR	Chiltern Railways (blue & white with a red stripe).
EM	East Midlands Trains {Connect} (blue with red & orange swish at unit ends).
FB	First Group dark blue.
FD	First Great Western & Hull Trains "Dynamic Lines" (dark blue with thin multi-coloured lines on lower bodyside).
FI	First Great Western "Local Lines" DMU (varying blue with local visitor attractions applied to the lower bodyside).
FS	First Group (indigo blue with pink & white stripes).
G	BR Southern Region/BR DMU green.
GA	Greater Anglia (white with red doors & black window surrounds).
GC	Grand Central (all over black with an orange stripe).
GW	Great Western Railway (TOC) dark green.
LM	London Midland (white/grey & green with black stripe around the windows).
LO	London Overground (all over white with a blue solebar & black window surrounds).
NO	Northern (deep blue, purple & white). Some units have area-specific promotional vinyls (see class heading for details).
NR	New Northern Rail (white & purple).
O	Non-standard (see class heading for details).
SN	Southern (white & dark green with light green semi-circles at one end of each vehicle. Light grey band at solebar level).
SR	ScotRail – Scotland's Railways (dark blue with Scottish Saltire flag & white/light blue flashes).
ST	Stagecoach {long-distance stock} (white & dark blue with dark blue window surrounds and red & orange swishes at unit ends).
SW	South Western Railway (two tone grey with a yellow lower bodyside stripe).
TP	TransPennine Express (silver, grey, blue & purple).
VI	Vivarail (white & green).
VT	Virgin Trains silver (silver, with black window surrounds, white cantrail stripe and red roof. Red swept down at unit ends).
XC	CrossCountry (two-tone silver with deep crimson ends & pink doors).

6.2. OWNER CODES

A   Angel Trains
AM   Alstom
BB   Balfour Beatty Rail Infrastructure Services
BN   Beacon Rail
BR   Brodie Leasing
CR   The Chiltern Railway Company
CS   Colas Rail
E   Eversholt Rail (UK)
GW   Great Western Railway (assets of the Greater Western franchise)
H1   Network Rail (High Speed)
HD   Hastings Diesels
MG   Mid Glamorgan County Council
NR   Network Rail
P   Porterbrook Leasing Company
RC   RailCare UK
SG   South Glamorgan County Council
SI   Speno International
SK   Swietelsky Babcock Rail
VI   Vivarail
VO   VolkerRail

6.3. OPERATOR CODES

AW   Arriva Trains Wales
CR   Chiltern Railways
EM   East Midlands Trains
GA   Greater Anglia
GC   Grand Central
GW   Great Western Railway
HD   Hastings Diesels
HT   Hull Trains
LM   London Midland
LO   London Overground
NO   Northern
SN   Southern (part of Govia Thameslink Railway)
SR   ScotRail
SW   South Western Railway
TP   TransPennine Express
VW   Virgin Trains West Coast
XC   CrossCountry

6.4. ALLOCATION & LOCATION CODES

Code	Depot	Depot Operator
AK	Ardwick (Manchester)	Siemens
AL	Aylesbury	Chiltern Railways
AN	Allerton (Liverpool)	Northern
CF	Cardiff Canton	Arriva Trains Wales/Colas Rail
CH	Chester	Alstom
CK	Corkerhill (Glasgow)	ScotRail
CZ	Central Rivers (Barton-under-Needwood)	Bombardier Transportation
DY	Derby Etches Park	East Midlands Trains
EX	Exeter	Great Western Railway
HA	Haymarket (Edinburgh)	ScotRail
HT	Heaton (Newcastle-upon-Tyne)	Northern
IS	Inverness	ScotRail
LM	Quinton Rail Technology Centre (Long Marston, Warwickshire)	Motorail Logistics
MN	Machynlleth	Arriva Trains Wales
NC	Norwich Crown Point	Greater Anglia
NH	Newton Heath (Manchester)	Northern
NL	Neville Hill (Leeds)	East Midlands Trains/Northern
NM	Nottingham Eastcroft	East Midlands Trains
OO	Old Oak Common HST	Great Western Railway
PM	St Philip's Marsh (Bristol)	Great Western Railway
RG	Reading	Great Western Railway
SA	Salisbury	South Western Railway
SE	St Leonards (Hastings)	St Leonards Railway Engineering
SJ	Stourbridge Junction	Parry People Movers
SU	Selhurst (Croydon)	Govia Thameslink Railway
TS	Tyseley (Birmingham)	London Midland
WN	Willesden (London)	London Overground
ZA	RTC Business Park (Derby)	Loram (UK)
ZB	Doncaster Works	Wabtec Rail
ZC	Crewe Works	Bombardier Transportation UK
ZD	Derby Works	Bombardier Transportation UK
ZG	Eastleigh Works	Arlington Fleet Services
ZH	Springburn Depot (Glasgow)	Knorr-Bremse Rail Systems (UK)
ZI	Ilford Works	Bombardier Transportation UK
ZJ	Stoke-on-Trent Works	Axiom Rail (Stoke)
ZK	Kilmarnock Caledonia Works	Wabtec Rail Scotland
ZM	Kilmarnock Bonnyton Works	Brodie Engineering
ZN	Wolverton Works	Knorr-Bremse Rail Systems (UK)
ZR	Holgate Works (York)	Network Rail

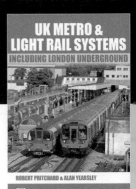